Geology of the

VOLUME TWO

The rugged northern coast of Ireland displays some of the most dramatic geology in the British Isles and the area has been of interest to geologists since the earliest days of the science. The close-packed columns of the Giant's Causeway, long renowned in folklore, and the ammonites in the 'Portrush Rock' were the subject of scientific debate and acrimony for over a century.

The first volume in this new edition of the Geological Survey Memoir describes the geology in terms which will be understood by the interested layman as well as the professional geologist. It includes sections on the relationship between geology and scenery, an account of human settlement in the area from Mesolithic times to the present day, a history of mining activity in the district and suggestions as to localities where the amateur collector may find minerals and fossils.

The second volume gives details of the stratigraphy, palaeontology and petrology of the rocks and logs of the deep boreholes. It also includes a comprehensive account of geophysical surveys of the area, which are of importance in assessing its potential for resources of hydrocarbons and geothermal energy.

GEOLOGICAL SURVEY OF NORTHERN IRELAND

Department of Commerce

H. E. WILSON and
P. I. MANNING

Geology of the Causeway Coast

Memoir for one-inch geological sheet 7

VOLUME TWO

CONTRIBUTORS

Stratigraphy and palaeontology
R. A. B. Bazley, T. P. Fletcher,
H. C. Ivimey-Cook, G. Warrington
and C. J. Wood

Petrology
J. R. Hawkes and R. W. Sanderson

Geophysics
J. R. P. Bennett, M. J. Bird
and J. Wheildon

Archaeology
A. E. P. Collins

BELFAST HER MAJESTY'S STATIONERY OFFICE 1978

© *Crown copyright 1978*

Bibliographical reference
WILSON, H. E. and MANNING, P. I. 1978.
Geology of the Causeway Coast. 2 vols.
Mem. Geol. Surv. North. Irel., Sheet 7.

Authors
H. E. WILSON, MSc, MIMM, MIGeol
*Geological Survey of Northern Ireland,
Belfast*

P. I. MANNING, TD, BSc, MIWES
*Institute of Geological Sciences,
Keyworth, Nottingham NG12 5GQ*

Contributors
R. A. B. Bazley, BSc, PhD and J. R. P. Bennett, BSc
*Geological Survey of Northern Ireland,
Belfast*

T. P. Fletcher, MSc, PhD and G. Warrington, BSc, PhD
*Institute of Geological Sciences,
Ring Road Halton, Leeds LS15 8TQ*

M. J. Bird, J. R. Hawkes, BSc, PhD,
H. C. Ivimey-Cook, BSc, PhD, R. W. Sanderson, BSc
and C. J. Wood, BSc
Institute of Geological Sciences, London

A. E. P. Collins, BA
*Archaeological Survey, Department of Finance
66 Balmoral Avenue, Belfast 9*

J. Wheildon, MSc, DIC
*Imperial College of Science and Technology,
South Kensington, London SW7*

ISBN 0 337 06094 0 – Vol 1
ISBN 0 337 06095 9 – Vol 2
ISBN 0 337 06096 7 – ⎡Vols 1 and 2⎤
⎣combined set⎦
Volume 2 is not sold separately

Publications of the Geological Survey of Northern Ireland

BOOKS

The composition and origin of the Antrim laterites
  and bauxites   62½p
The geology of the country around Dungannon   £1.50
The geology of the country around Ballycastle   £2.00
The geology of Belfast and the Lagan Valley   £2.00
Sources of aggregate in Northern Ireland
  (2nd edition)   70p
The use and resources of moulding sand in
  Northern Ireland   30p
The regional geology of Northern Ireland   50p

*Obtainable from Government Bookshops in Belfast, London,
Edinburgh, Cardiff, Manchester, Birmingham and Bristol,
or any bookseller. Post orders to Government Bookshop,
Chichester Street, Belfast, BT1 4JY.
Prices do not include postage*

GEOLOGICAL MAPS

Six inches to one mile (1:10 560) maps of coalfield areas:
  Antrim: 5 SW, 5 SE, 9 NW: Tyrone: 46 SE, 47 NW,
  47 SW, 54 NE   60p *each*

Three inches to one mile (1:21 120) map:
  Engineering geology of the Belfast district   £1.50

One inch to one mile (1:63 360) colour-printed maps:
  Giant's Causeway (7) Sheet (Solid)   80p
  Giant's Causeway (7) Sheet (Drift)   80p
  Ballycastle (8) Sheet (Solid)   80p
  Ballycastle (8) Sheet (Drift)   80p
  Carrickfergus (29) Sheet (Solid)   80p
  Carrickfergus (29) Sheet (Drift)   80p
  Pomeroy (34) Sheet (Solid)   (*in press*)
  Pomeroy (34) Sheet (Drift)   (*in press*)
  Dungannon (35) Sheet (Solid)   80p
  Dungannon (35) Sheet (Drift)   80p
  Mourne Mountains Special Sheet (Solid)   (*in press*)

1:250 000 map:
  Northern Ireland   £2.00

GEOPHYSICAL MAPS

Quarter-inch to one mile (1:253 440) maps:
  Gravity anomaly map of Northern Ireland   75p
  Aeromagnetic map of Northern Ireland   75p

*Obtainable from Ordnance Survey, Ladas Drive, Belfast,
BT7 9FJ, or Ordnance Survey agents*

*A full list of provisional maps and open-file reports may be
obtained from the Geological Survey of Northern Ireland,
20 College Gardens, Belfast, BT2 6BS*

CONTENTS

VOLUME ONE

Part One Geology of the district

2 **Chapter 1 Introduction**
Geological history 2
Previous literature 4

6 **Chapter 2 Permo-Triassic rocks**
Permian 6
Triassic 7
  Sherwood Sandstone Group 7
  Mercia Mudstone Group 7
  Rhaetic 8

11 **Chapter 3 Jurassic rocks**
Portrush 11
White Park 12
Port More Borehole 13
Records of Liassic specimens from later deposits 14

15 **Chapter 4 Cretaceous rocks**

23 **Chapter 5 Tertiary events**
Tertiary igneous activity 23
Sequence of igneous events 23
Lavas and tuffs 24
  Lower Basalts 24
  Interbasaltic Bed 26
  Middle (Tholeiitic) Basalts 26
  Upper Interbasaltic Bed 28
  Upper Basalts 29
Intrusive igneous rocks 29
  Volcanic vents 29
  Volcanic plugs 38
  Sills 39
  Dykes 40
The post-volcanic period 41

42 **Chapter 6 Structure**

43 **Chapter 7 Pleistocene**
Glacial striae and roches moutonnées 44
Boulder clay 44
Glacial sand and gravel 44
Glacial drainage channels 45

46 **Chapter 8 Recent**
Raised beaches 46
Submerged forest 47
Estuarine clay 48
Blown sand 48
Present-day marine and estuarine deposits 49
River terraces 49
Alluvium 49
Peat 50

51 **Chapter 9 Economic geology**
Iron ore; bauxite; chalk; basalt; lignite; peat; water

Part Two Geology and Man

54 **Chapter 10** Geology and scenery

58 **Chapter 11** Human occupation and industry

60 **Chapter 12** Minerals and mines

61 **Bibliography**

68 **Appendix 1** Localities for collectors

69 **Appendix 2** Glossary of geological terms

71 **Index** (to Volume One)

VOLUME TWO

Part Three Details of the geology

75 **Chapter 13 Triassic rocks**
Formations of the Mercia Mudstone Group 75
  Collin Glen Formation 75
  Port More Formation 75
  Knocksoghey Formation 75
  Glenstaghey Formation 75
  Craiganee Formation 76
  Lagavarra Formation 76
  Palynology of the Triassic Sequence in the Port More Borehole 76
  Correlation 77

78 **Chapter 14 Jurassic rocks**
Lower Lias 78
Stratigraphical palaeontology of the Lower Jurassic of the Port More Borehole 80

84 **Chapter 15 Cretaceous rocks**
Stratigraphy 84
Detailed stratigraphical palaeontology 90
  The Hibernian Greensands Formation 90
  The White Limestone or Chalk Formation 90
    The pre-North Antrim Hardground succession: Galboly Chalk; Cloghastucan Chalk; Creggan Chalk; Boheeshane Chalk; Larry Bane Chalk; Ballintoy Chalk; Glenarm Chalk 90
    The post-North Antrim Hardground succession: Glenarm Chalk; Garron Chalk; Portrush Chalk; Ballymagarry Chalk; Tanderagee Chalk 99
    The post-Tanderagee Chalk succession 111
    The White Limestone of the Port More Borehole 111
Stratigraphical range charts 113

116 **Chapter 16 Tertiary extrusive igneous rocks**
    Details of exposures   116
      Lower Basalts   116
      Interbasaltic Bed   119
      Tholeiitic Basalts   124
      Upper Interbasaltic Bed   126
      Upper Basalts   127
    Petrology of the extrusive igneous rocks   130
      Lower and Upper Basalts   130
      Tholeiitic Basalts   131
      Pyroclastic rocks   131
      Interbasaltic Beds   132

133 **Chapter 17 Intrusive igneous rocks**
    Volcanic vents   133
      The Carrickarade Vent   133
      Minor vents   134
    Volcanic plugs   135
    Sills   135
      Portrush Sill   135
      Knocksoghey Sill   137
    Dykes   138
    Petrology of the intrusive igneous rocks   139
      Vents   139
      Plugs   139
      Sills   139
      Dykes   142

143 **Chapter 18 Geophysical investigations**
    Gravity surveys   143
    Magnetic surveys   144
    Seismic surveys   145
    Port More Borehole   148
      Core analysis   148
      Geophysical logging   150
      Heat-flow measurement   155
      Conclusions   156

157 **Bibliography**

163 **Appendix 3**   Records of boreholes
    Corbally Reservoir Borehole   163
    Port More Borehole   163

167 **Appendix 4**   Chemical analyses of rocks

168 **Appendix 5**   Geological Survey photographs

170 **Index**

PLATES

1   Giant's Harp and the Chimney Tops, Giant's Causeway   2
2   'Portrush Rock', Lias shales hornfelsed by the intrusion of the Portrush Sill   9
3   'Portrush Rock' with ammonites   9
4   White Park Bay   10
5   Elephant Rock—sea-weathered White Limestone   19
6   Ballintoy Port   19
7   Bengore Head   20
8   Compound lava flow, Dunluce   20
9   Giant's Causeway from the air   21
10   The Grand Causeway   22
11   Spaniard Rock and the Chimney Tops   22
12   Columnar basalt, Grand Causeway   31
13   Columnar basalt, Ballynastraid Quarry   31
14   Spheroidal weathering of basalt   32
15   Relict spheroids in the Interbasaltic Bed   32
16   Ballintoy from the air   33
17   Carrickarade Island   34
18   Carrickarade   34
19   Explosion vent, Devil's Port   35
20   Volcanic breccia, White Rocks   35
21   Portrush and Ramore Head   39
22   Layered sill, Large Skerries   37
23   Basalt dyke, Portnaboe   37

24   Triassic miospores from the Glenstaghey Formation, Port More Borehole   101
25   The Giant's Cut   102
26   Basal members of the White Limestone, Oweynamuck   102
27   Chalk cliffs, west end of White Park Bay   103
28   Chalk cliffs near Portbraddan   103
29   Boheeshane Chalk, Eastern Spring, White Park Bay   104
30   Cliff section: Larry Bane Head, west side   104
31   Aerial view of Chalk cliffs, White Park Bay   105
32   Chalk cliffs, east side of Larry Bane Head   105
33   Cliff west of Larry Bane Head   106
34   Larry Bane Head and Quarry from the air   107
35   Boheeshane Bay cliffs from the air   107
36   Chalk cliffs west of Dunluce   108
37   White Rocks and Ballymagarry Quarry from the air   108

FIGURES

1 Geological formations present in the district 2
2 Sketch-map of the geology of the Causeway Coast 3
3 Zones and local ranges of the Lower Lias in North Antrim 10
4 Sketch-map of the Portrush Sill, showing outcrops of hornfelsed Lias mudstone 12
5 Outcrops of Liassic rocks at White Park Bay 13
6 Subdivisions of the White Limestone 15
7 Sketch-map of Ballintoy Harbour 25
8 Sketch-section across Ballintoy Harbour 25
9 Diagrammatic section of the coast from Roveran Valley Head to Weir's Snout 27
10 Sketch-map of the Giant's Causeway 28
11 Locations of intrusive igneous features along the Causeway Coast 29
12 Sketch-map of Carrickarade 30
13 Ice-margin deposits at Portballintrae 43
14 Section along the coast from Portrush to Carrickarade 54

15 Formations of the Triassic rocks in the Port More Borehole 74
16 Ammonite ranges from the Jurassic of the Port More Borehole 81
17 Summarised stratigraphy of the White Limestone 85
18 Composite lithostratigraphic succession of the White Limestone 86–87
19 Variation in the Lower Campanian succession from Boheeshane to Portbraddan 88
20 Locations of important sections in the coastal outcrop of the Cretaceous rocks 89
21 Port More: composite cliff and platform succession 109
22 Port More Borehole: graphic sections of the Cretaceous rocks 112
23 Vertical distribution of selected species in the upper part of pre-North Antrim Hardground succession 114
24 Vertical distribution of selected species in post-North Antrim Hardground succession 115
25 Locations of old mines in the Interbasaltic beds in the area south of Portrush 120
26 Locations of old shafts, adits and trial pits in the Ballintoy–White Park area 122
27 Map showing Bouguer gravity values at sea level 124
28 Map showing total force magnetic anomalies from aeromagnetic survey 146–147
29 Graph showing relationship between dry bulk density and effective porosity of the Sherwood Sandstone Group 149
30 Geophysical logs of the Port More Borehole 151–153
31 Correlation of borehole temperature measurement, laboratory measurement of conductivity, heat-flow calculation and geology of the Port More Borehole above 600 m 156

TABLES

1 Summary of the Lower Jurassic of the Port More Borehole 14
2 Macrofauna of the Sinemurian Lias, White Park Bay 79
3 Chronostratigraphic divisions identified in the Lower Jurassic of the Port More Borehole 80
4 Laboratory determinations of density, porosity, permeability and isotropy of Sherwood Sandstone Group sandstones of the Port More Borehole 149
5 Laboratory determinations of density, porosity, and sonic velocity in some Triassic rocks of the Port More Borehole 154

PART THREE

Details of the geology

74 CHAPTER 13 TRIASSIC ROCKS

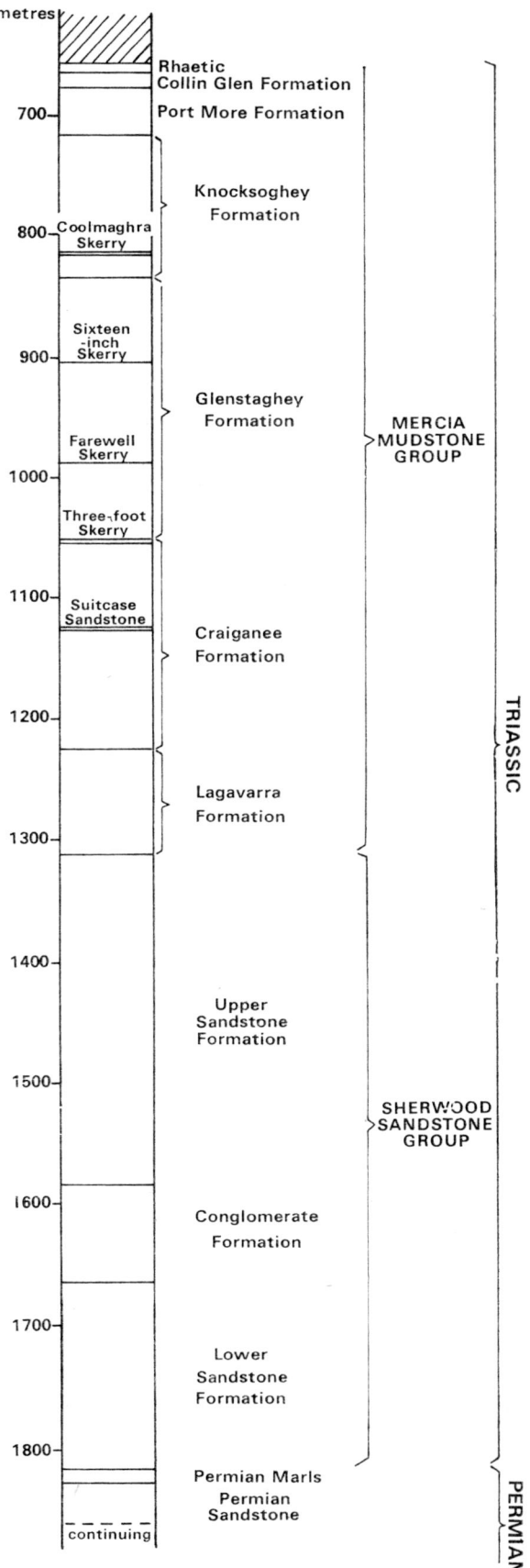

Figure 15 Formations of the Triassic Rocks in the Port More Borehole

CHAPTER 13
Triassic rocks

FORMATIONS OF THE MERCIA MUDSTONE GROUP

With the exceptions of the Rhaetic and the highest (Collin Glen) formation of this group, Triassic rocks seen at outcrop in Ulster cannot be precisely correlated. For the definition of the majority of the formations within the Mercia Mudstone Group, therefore, borehole sections must be used as stratotypes. The Port More Borehole section (Figure 15) is here adopted as the stratotype for formations of the Mercia Mudstone Group, below the Collin Glen Formation, in Northern Ireland. The Port More sequence is divisible into the following formations which have the thicknesses indicated and are present in the borehole in the depth range shown. Discrepancies between thickness and borehole intercepts are accounted for by the presence of igneous intrusions, which are not included in the former.

Collin Glen Formation
About 10.82 m thick (664.77–675.59 m in borehole).
Green calcareous mudstones

Type locality: Collin Glen [J 272 717] (Manning and others, 1970, p. 41). In the Port More section this formation consists of greenish-grey calcareous mudstones, massive and block-jointed in the upper part but with thin laminae and beds of paler cementstones in the lower. The cementstones show contemporaneous brecciation. Because of the loss of 1.5 m of core at the top of the formation the junction with the Rhaetic was not seen but would appear to be abrupt with a sharp lithological change to the black shales of the Lower Rhaetic with their assemblage of small bivalves. *Euestheria sp.* aff. *portlocki* (Jones) occurs between 666.29 and 666.75 m. Towards the base some red-brown mudstone occurs, but the base is drawn below the greenish massive mudstone at 675.59 m. The total thickness compares closely with that of the same formation in the Larne Borehole, 53 km to the southeast, and this formation is doubtless the equivalent of the 'Tea Green Marl' in the Parva Formation (Elliott, 1961) in England.

Port More Formation
38.07 m thick (675.59–713.66 m).
Red-brown poorly bedded mudstones without anhydrite nodules or gypsum

The uppermost red beds of the Mercia Mudstone Group consist of dull or drab brown mudstones with some greenish mottling at a number of levels. These pass down into mudstones of a brighter red-brown colour. Between 697.99 and 703.17 m small quartz pebbles and grains in a red-brown silty mudstone matrix were recorded while some quartz grains occurred below this. A few thin green beds were also present towards the base. The base was drawn immediately above mudstones containing nodular anhydrite. At Magilligan and Larne, as well as in boreholes in England, a similar zone of mudstones without gypsum or anhydrite typically occurs beneath the Collin Glen Formation and its lithostratigraphic correlatives.

Knocksoghey Formation
122.64 m thick (713.66–836.30 m).
Red-brown poorly bedded mudstone with gypsum and nodular anhydrite

Monotonous red-brown mudstones, relatively unbedded and frequently silty, are dominant in the Knocksoghey Formation. In the top 43.03 m large anhydrite nodules up to 0.17 m across and of a characteristic bluish white colour are common. Green reduction spots (with some 'fish-eyes') are abundant. Between 756.65 and 778.31 m there are occasional greenish mudstone layers and the anhydrite nodules are smaller and frequently subangular, or anhydrite may be disseminated throughout the mudstone. Some fibrous gypsum veins occur.

For a further 23.19 m below this anhydrite nodules are numerous and are often of a bluish-white colour. The mudstones remain red-brown in colour with green reduction spots and some mottling while quartz grains were noted at some levels.

Below 801.50 m the anhydrite is usually well disseminated, frequently as specks and patches rather than in nodular form. The mudstones are silty at some levels and a pale-green siltstone 48 cm thick occurred at 813.13 m. Rounded sand grains were observed in the mudstone above this siltstone.

The base of the Knocksoghey Formation is taken above the uppermost well-bedded mudstone of the underlying Glenstaghey Formation which is further distinguished by the presence of fine sandstones, more numerous siltstones, a noticeable colour banding of laminated mudstones and fewer fish-eyes and anhydrite nodules than the overlying formation.

The Coolmaghra Skerry (815.24–817.78 m) is a greenish to buff dolomitic sandstone member with thin mudstone partings and some anhydrite.

Glenstaghey Formation
187.2 m thick (836.3–1056.1 m).
Laminated and rhythmically bedded mudstones and siltstones

The formation comprises predominantly bedded mudstones with subordinate siltstone and some fine sandstones; the colour ranges from red-brown, through chocolate to grey and shades of green. Green beds are more common than elsewhere in the group except in the Collin Glen Formation.

Anhydrite nodules are rarer than in the Knocksoghey Formation and thin gypsum veins are plentiful. At 857.33 m darker chocolate-brown mudstones are present for the first time. Dark blue-grey beds are common. Fault gouges were recorded between 838.12 and 838.63 m and between 838.81 and 839.62 m. In the remainder of the sequence the mudstones range from red-brown to dark chocolate-brown with dark green laminations and subordinate thin greenish siltstone bands in rhythmic sequences. Some flow brecciation was noted. The Sixteen-Inch Skerry Member, 41 cm thick at 908 m, is an off-white to pale-green sandstone with reddish silty banding. Small anhydrite patches are common and small-scale microfaulting was present.

Below about 920 m the mudstones have more numerous silty laminae giving the rock a distinctive 'striped' appearance. Some bands exhibit polygonal mudcracks infilled by the immediately overlying sediment. A medium to coarse-grained intrusive dolerite occurred between 932.18 and 960.12 m, metamorphosing the formation both above and below. Below the intrusion red-brown more massive mudstones alternate with laminated chocolate-brown and dark-green mudstones. A few thin silty sandstone layers occur towards the Farewell Skerry Member, a micaceous sandstone, off-white to pale-green, 0.89 m thick at 989.6 m. Below 1009.75 m a 1.55-m thick grey-green laminated mudstone unit with numerous gypsum bands is underlain by red-brown mudstones with grey-green layers, often well laminated. There are occasional autobrecciated horizons and gypsum preudomorphs after halite at some levels.

Small bluish-white anhydrite nodules reappear below 1033.78 m. A dolerite intrusion occurs between 1035.69 and 1040.48 m with some low-grade metamorphism of the mudstones. Below the intrusion the red-brown mudstones are commonly silty and include rare thin silty sandstone layers. Rhythmic sequences are common. Between 1049.22 and 1050.52 m the mudstone sequence is well laminated in shades of bright red-brown and chocolate-brown with occasional green mudstone layers and thin siltstone bands. The latter are often broken by small mudstone-infilled cracks.

The base of the formation is taken above the Three-foot Skerry below which siltstones and thin sandstone beds become more important in the underlying formation.

Craiganee Formation
166.8 m thick (1056.1–1222.9 m).
Laminated mudstones and siltstones with thin pale sandstones

The Craiganee Formation is broadly similar to the overlying Glenstaghey Formation but contains a higher proportion of siltstone beds and the mudstone units are less well laminated. The number and individual thickness of silty and sandy beds increases towards the base of the formation where there is a noticeable increase in the thickness of sandstone beds and their importance in the underlying Lagavarra Formation.

The Three-foot Skerry (actually 1.07 m thick) is an off-white sandstone with a silty top and the sequence below it includes numerous similar though thinner (average about 50 cm) sandstones interbedded with red-brown mudstones and siltstones in rhythmic sequences. Mudcracks, sand volcanoes, ripple marks and mudstone breccias and anhydrite, in specks and small nodules, are common throughout. At 1122.5 m there is a 'skerry' 1.4 m thick, of pale silty sandstone, locally displaying current bedding and slump structures.

The Suitcase Sandstone, 0.9 m thick at 1127.50 m, is a buff-coloured faintly colour-banded calcareous fine-grained sandstone.

Lagavarra Formation
94.4 m thick (1222.9–1317.3 m)

This formation consists of red-brown, silty, micaceous mudstones, chocolate-brown mudstones, red-brown siltstones and buff to off-white sandstones in well developed rhythmic sequences though with sandstone increasing in importance towards the base of the formation. The mudstones often have well developed microcyclothems. There are, for example, 17 microcyclothems between 1141.78 and 1144.40 m, each unit, between 4 and 25 cm thick, showing a sharp base, a red-brown mudstone bottom and a thin pale silty top, sometimes showing cracking or pull-apart structures. The sandstones exhibit load-casts and flow structures at some levels and the mudstones are autobrecciated at some horizons. Numerous spherical calcareous patches characterise a sandstone unit at 1171.24 m. Some anhydrite in the form of bands of small nodules is present in small amounts and some halite pseudomorphs were observed. The base was drawn at the top of the highest thick sandstone of the Sherwood Sandstone Group and below the lowest recorded pseudomorph after halite and the lowest thick and fairly prominent siltstone and mudstone beds intercalated with sandstones.   PIM

PALYNOLOGY OF THE TRIASSIC SEQUENCE IN THE PORT MORE BOREHOLE

A total of thirty samples, comprising eight from the Rhaetic and twenty-two from the Mercia Mudstone Group have been processed and examined for palynological residues. All the Rhaetic material was found to be barren, a disappointing result in view of the prolific nature of microfloras obtained from this unit in the Langford Lodge (see Warrington *in* Manning and others, 1970) and Larne boreholes (Manning, Wilson and others, 1975.) The absence of miospores in the Port More Rhaetic samples is considered to result from thermal metamorphism associated with an intrusive dolerite body which occurs above the 4.0 m of sediments assigned to the Rhaetic in the borehole.

The samples examined from the Mercia Mudstone Group originated from four of the six formations of that group as follows:

Collin Glen Formation	three samples
Port More Formation	not sampled
Knocksoghey Formation	three samples
Glenstaghey Formation	twelve samples
Craiganee Formation	four samples
Lagavarra Formation	not sampled

The Sherwood Sandstone Group below the Mercia Mudstone, was not sampled. Only four samples, all from the Glenstaghey Mudstone Formation, yielded miospores (Plate 24). The composition of the assemblages obtained is as follows:

Sample **SAL 782** (depth 842.47 m)
Alisporites cf. *circulicorpus* Clarke 1965
A. cf. *grauvogeli* Klaus 1964
A. cf. *parvus* de Jersey 1962
A. sp.
?Scopulisporites minor Mädler 1964
Sulcatisporites sp.
Brachysaccus neomundanus (Leschik) Mädler 1964
Protodiploxypinus cf. *decus* Scheuring 1970
?Triadispora plicata Klaus 1964
?T. stabilis Scheuring 1970
T. sp.
Ellipsovelatisporites plicatus Klaus 1960
?Illinites chitonoides Klaus 1964
Parillinites cf. *vanus* Scheuring 1970
?Striatoabieites aytugii Visscher emend. Scheuring 1970
Lunatisporites sp. sensu Visscher 1966
Ovalipollis sp.
Cycadopites accerrimus (Leschik) Clarke 1965
C. sp.
Retisulcites perforatus (Mädler) Scheuring 1970
Echinitosporites iliacoides Schulz & Krutzsch 1961
Circumpolles (indeterminate; rare)
Bisaccate spp. (indeterminate; numerous but commonly damaged or with a very dark exine, particularly in the case of large specimens)

Sample **SAL 785** (depth 898.25 m)
Monosulcites sp.
Bisaccate spp. (indeterminate)

Sample **SAL 788** (depth 966.23 m)
Punctatisporites sp.
Alisporites grauvogeli
A. sp.
Scopulisporites minor
Sulcatisporites kraeuseli Mädler 1964

?*Triadispora crassa* Klaus 1964
T. plicata
cf. *T. plicata*
Angustisulcites klausii Freudenthal 1964
Illinites chitonoides
?*I. kosankei* Klaus 1964
?*Striatoabieites aytugii*
Tsugaepollenites oriens Klaus 1964
Bisaccate spp. (indeterminate)

Sample **SAL 792** (depth 999.76 m)
Punctatisporites sp.
Alisporites cf. *toralis* (Leschik) Clarke 1965
A. sp.
cf. *Colpectopollis ellipsoideus* Visscher 1966
Sulcatisporites kraeuseli
cf. *Voltziaceaesporites heteromorpha* Klaus 1964
Bisaccate spp. (indeterminate)

Miospores were abundant only in sample SAL 782; the other samples contained only sporadic or, in the case of SAL 785, very infrequent specimens. The miospores in each assemblage were only moderately well preserved, many specimens being very dark coloured or showing signs of damage. Thermal effects associated with two dolerite intrusions in the Glenstaghey Mudstone Formation may have caused some of the darkening noted in the specimens. One assemblage (from SAL 788) comes from clearly metamorphosed sediments 6.11 m from the higher of the intrusions and two others (SAL 785 and 792) are from within 30 and 40 m respectively of this dolerite; SAL 782 originated over 80 m from the same intrusion.

Age of the miospore assemblages

SAL 782: The assemblage from this sample contains several stratigraphically useful taxa including *Echinitosporites iliacoides* and *Retisulcites perforatus* which have both been found to occur sparingly in the Lettenkohle but fairly abundantly in the succeeding lower part of the Gipskeuper in sections studied by Scheuring (1970) near Basle. The miospores *Protodiploxypinus decus* and *Parillinites vanus* (specimens comparable with which have been recognised in the Port More assemblage) were found by Scheuring to occur in the lower part of the Gipskeuper only and to range about as high in the sequence as *E. iliacoides* and *R. perforatus*. The presence of the taxa referred to above indicates that the horizon of the SAL 782 assemblage is correlatable with the lower Gipskeuper—possibly the part of the sequence delimited by the palynological subdivisions upper A, B and lower C in Scheuring's scheme (1970, p. 107)—and may be regarded as early Carnian in age. The majority of the remaining miospores observed in the assemblage are of little precise stratigraphical value although the presence of ?*Triadispora stabilis* and *Ellipsovelatisporites plicatus* supports the suggestion of a Carnian age.

SAL 785: The remains recorded from this sample provide no evidence of age.

SAL 788: The presence of *Angustisulcites klausii* (known from the Upper Bunter and Lower Muschelkalk (Visscher, 1966; Visscher and Commissaris, 1968), *Illinites chitonoides* (recorded from the Muschelkalk and Lettenkohle; Geiger and Hopping, 1968), ?*I. kosankei* (known from the Upper Bunter and Lower Muschelkalk; Geiger and Hopping, 1968) and *Tsugaepollenites oriens* (known from the Middle Muschelkalk; Klaus, 1964) indicates that the horizon of sample SAL 788 is probably equivalent to part of the Muschelkalk of Europe. The deposit from which the sample was obtained is, therefore, probably of Anisian or earliest Ladinian age.

SAL 792: Specimens obtained from this sample are indicative of no more than a late-Early or Middle Triassic age.

In summary, the miospore evidence indicates that, in the Port More section, the Glenstaghey Formation at a depth of 842.47 m is of early Carnian age, while 123.76 m lower at 966.23 m, an Anisian or early Ladinian age is likely. The sporadic miospores obtained at 999.76 m suggest a late-Early Triassic or Middle Triassic age.

Correlation

Although the Mercia Mudstone Group in the Port More Borehole has yielded only a small number of generally meagre palynological assemblages, sufficient material has been obtained to indicate certain general correlations with other sections which have been studied palynologically. In Northern Ireland, the Mercia Mudstone Group has been studied in the Langford Lodge Borehole (Warrington *in* Manning and others, 1970), some 68 km south of the Port More borehole site. Only rock-cutting samples were available from this borehole but some 155 m of strata occurring below the Rhaetic (from about 823 m to about 978 m in the borehole section) were regarded as equivalent to the German Lower and Middle Keuper and were assigned ages of Ladinian to Norian. The highest productive sample from Port More (assigned an early Carnian age) indicates a correlation between an horizon in the lower part of this 155-m thick unit at Langford Lodge and the strata occurring 175.70 m below the Rhaetic at Port More. Thus, the approximately coeval strata appear to be slightly thicker at Port More though the absence of evidence on which to define the base of Ladinian strata in that section renders a precise comparison impossible.

The miospore assemblage from a depth of 966.23 m (about 300 m below the Rhaetic) in the Port More Borehole contains *Tsugaepollenites oriens* which has been recorded at about 201 and 231 m below the Rhaetic in the Langford Lodge section thus indicating a general correlation between strata at those levels in the two sections and again evidencing a slightly greater thickness of strata at Port More.

Sections in Institute of Geological Sciences boreholes near Blackpool have been studied recently (Evans and Wilson, 1975) and represent horizons up to approximately the same stratigraphic level as the highest productive sample from Port More. The Blackpool sections have yielded *Tsugaepollenites oriens* from the Kirkham Mudstone Formation at horizons both slightly above and below the Preesall Salt (Warrington, 1974) and thus a general correlation may be established between that part of the sequence in west Lancashire and part of the Glenstaghey Mudstone Formation of the Port More section and the Mercia Mudstone Group (undifferentiated) of the Langford Lodge Borehole. The Breckells Mudstone Formation, overlying the Kirkham Mudstone Formation, has yielded miospore assemblages containing *Echinitosporites iliacoides* and *Retisulcites perforatus* in the Blackpool area (Warrington, 1975). A correlation between part of the Breckells Mudstone of west Lancashire and the horizon of the highest productive sample from Port More (from the topmost part of the Glenstaghey Formation) is therefore indicated.

Any correlation, other than in the most general terms, is impossible on the basis of the lowest assemblage from Port More.

GW

CHAPTER 14

Jurassic rocks

Lower Lias

As described in Part One, only the Sinemurian and Lower Pliensbachian zones from *semicostatum* to *ibex* have been recognised in this area, including zones proved in the deep borehole at Port More (Table 3). Details of exposures and faunas are given below.

At the western end of White Park Bay (Figure 5) a small stream (Lemnagh Burn) reveals small disconnected exposures of Lias shales which may be slipped masses. From this stream section, Loc. 1, [D 0155 4378], 800 m on a bearing N291° from Lemnagh Beg House, a fauna indicative of the *Echioceras raricostatum* Zone was obtained. Among the ammonites were: *Crucilobiceras sp.*, *Gemmellaroceras* [*Tubellites*] *tubellus* (Simpson), *Leptechioceras sp.*, *Paltechioceras boehmi* (Hug) and *Paltechioceras sp.* The *Leptechioceras* indicates the *Leptechioceras macdonnelli* Subzone. The belemnite *Nannobelus acutus* (Miller) was also recorded. Formerly, Tate (1867, p. 304) used this as a 'zone fossil' of a Zone of '*Belemnites*' *acutus*. The remaining fauna is listed, with that of the other localities yielding fossils attributed to the *raricostatum* Zone, in Table 2.

At the base of the small waterfall, Loc. 2 [D 0280 4426], 360 m on a bearing N285° from Ballintoy School, towards the eastern end of White Park Bay, the basal conglomerate of the Chalk rests on 0.6 m of dark-grey Lias shales containing a few calcareous nodules (Figures 5 and 19). The fauna of the shales probably indicates the *raricostatum* Zone. Ammonites recorded include *Gemmellaroceras tubellus* (Simpson) and *Gemmellaroceras sp.*

Lias shales were formerly exposed in small stream sections amongst the landslips to the north of this exposure, close to the shore at Loc. 4, [D 0261 4446], 600 m on a bearing N298° from Ballintoy School; an exposure in the bed of this small stream yielded a *raricostatum* Zone fauna. Ammonites included *Cheltonia sp.*, '*Hemimicroceras*' cf. *cingendum* Trueman & Williams, *H.* cf. *vitreum* (Simpson) and *H. sp.*

To the south-west at Loc. 3 [D 0230 4426], mudstones temporarily exposed amongst the sand of the beach yielded an ammonite fauna including *Eoderoceras sp.*

The small promontory at Oweynamuck at the eastern end of White Park Bay contains small exposures of slipped Liassic shale beneath the Cretaceous. Three small exposures on the south-west of the point yielded faunas which could belong to the *raricostatum* Zone. These faunas are listed as localities 5–7 in Table 2. At Loc. 5 [D 0288 4484], the combination of *Gemmellaroceras tubellum* and *Oxynoticeras sp.* juv. with the bivalve fauna suggests the *raricostatum* Zone, but this is not conclusive since both ammonites range up into the *jamesoni* Zone.

Two separate small exposures of grey silty shales and mudstones occur in landslips on the south-west side of Oweynamuck. The exposures are mainly intertidal and dip steeply into the landslips. Locality 6 [D 0288 4481] is 1006 m on a bearing N270° from the church above Ballintoy Harbour and 640 m on a bearing N335° from Ballintoy School. About 9 m of steeply dipping and poorly exposed Liassic mudstones were seen in the beach. Within this sequence the top 1.3 m was of particularly dark and micaceous mudstone; these beds overlie more massive mudstone, locally silty and with occasional calcareous nodules. The only ammonite found was an indeterminate deroceratid. About 45 m to the north (Locality 7) about 3 m of Lias was seen below the phosphatic nodule bed forming the base of the Cretaceous. The dark-grey silty mudstones dip obliquely to those in the other exposure and appear to be part of a separate landslip. No ammonites were found but the fauna is consistent with a *raricostatum* Zone age.

On the western extremity of Oweynamuck [D 0285 4484] an exposure of Liassic mudstone was collected in 1956 and yielded a rich fauna including: terebratuloid fragments, *Procerithium sp.*, *Zygopleura berthaudi* (Dumortier), *Camptonectes mundus* Melville, *C. sp.*, *Cardinia attenuata* (Stutchbury), *Cercomya sp.*, *Chlamys rollei* (Stoliczka), *Entolium liasianum* (Nyst), *Grammatodon insons* Melville, *G. sp.*, *Gryphaea sp.*, *Liostrea hisingeri* (Nilsson), *Mactromya arenacea* (Terquem), *Modiolus hillanoides* (Chapuis & Dewalque), *Nuculana* (*Rollieria*) *bronni* (Andler), *Oxytoma inequivalve* (J. Sowerby), *Palaeoneilo galatea* (d'Orbigny), *Pteromya tatei* (Richardson & Tutcher), *Plicatula hettangiensis?* Terquem, *P. sp.*, *Protocardia truncata* (J. de C. Sowerby), *Pseudolimea acuticostata* (Münster), *Ryderia doris* (d'Orbigny), *Aegoceras* (*Beaniceras*) *centaurus* (d'Orbigny), *Cheltonia?*, *Tropidoceras?*, and undetermined belemnites. The ammonites show that these beds belong to the *Tragophylloceras ibex* Zone, *valdani* Subzone, which is also known from a small exposure at Portnakillew, west of Ballycastle, and from the Port More Borehole.   HC I-C

The Lias probably also underlies a small patch of beach between the eastern end of White Park Bay and Ballintoy harbour. The base of the cliff of White Limestone shows an horizon only a few metres above the base of the Cretaceous. No Lias was seen in the floor of the old Ballintoy Quarry [D 037 452] where Tomkeieff and Patterson (1953) mapped a small area, but the quarry floor of White Limestone is probably only a metre or so above the Cretaceous base. Similarly, the western half of Boheeshane Bay may well be floored by Lias shales.   PIM

From the exposures in the White Park Bay area where the Cretaceous is seen resting directly on the Lower Lias, it would appear that, while there is no observable discordance, the Chalk may transgress onto lower horizons to the west. There is no indication of higher zones which may have been formerly deposited during the Lower Lias–Cretaceous interval.

Faunal lists of Liassic material from White Park Bay have been given by Charlesworth and Preston (1960) and by Charlesworth (1963, p. 346, 348). These lists, like the earlier one of Portlock (1843), contain ammonites whose identification suggests the presence of zones as low as that of *Psiloceras planorbis* and also the presence of the *Prodactylioceras davoei* Zone at the top of the Lower Lias. No ammonites below those of the *raricostatum* Zone or above the *ibex* Zone were collected on the present resurvey and the gentle dome structure suggests that if outcrops of lower zones occur, they would be confined to an area lying beneath the landslip or blown sands. It is doubtful whether the fully developed zonal sequences hitherto claimed can be substantiated. It is possible that earlier lists may contain specimens not actually collected in situ. Tate (1867, p. 304) remarks that the majority of fossils, principally ammonites enclosed in calcareous nodules, which were on sale at the Giant's Causeway, were obtained during the winter on the shore at Ballintoy. Specimens obtained from other areas in the north of Londonderry, the east coast of Antrim or from the Drift, could account for the anomalous records.

Remanié Lower Lias fossils found in the basal Cretaceous conglomerate at Oweynamuck include *Coroniceras sp.* from the *Arietites bucklandi* Zone, '*Hemimicroceras*' and *Paltechioceras* aff. *favrei* (Hug) from the *raricostatum* Zone, also *Polymorphites costatus*

(Quenstedt) and *Uptonia?* from the *Uptonia jamesoni* Zone. *Uptonia sp.* was also found from the conglomerate at [D 0280 4426].

Derived Liassic ammonites in the basal conglomerate of the Cretaceous have also been recorded from other localities. Savage (1963, pp. 179–180) recorded a *Dactylioceras sp.* from this bed at Murlough Bay [D 193 417] and revised the determination of the same genus given from another specimen by Hartley (1933, pp. 238–239). These ammonites indicate erosion of Toarcian sediments. Versey (1958, p. 440) recorded a specimen of *Pleuroceras transiens* (Frentzen) from a roadside above Binvane Farm, Murlough Bay. This species is known from a few metres of strata at the boundary of the *margaritatus* and *spinatum* zones of the Middle Lias in Raasay.

The Carrickarade vent agglomerate has yielded fragments of Liassic rocks (Tomkeieff and Patterson, 1947, p. 89) and these included a *Polymorphites?* fragment suggesting the *jamesoni* Zone (Manning in Patterson, 1962, p. 71) together with longer-range bivalves. (See Chapter 3).

Both Lower and Middle Lias remanié fossils have been discovered in the drift along the north coast in a number of places including Ballintoy and Portrush (Langtry, 1875; Tate, 1870). Charlesworth (1963, p. 349) gives the following list of Lias fossils: '*Cincta numismalis, Entolium liasianum, Eotrapezium cucullatum, Oxytoma inaequivalve, Pseudolimea acuticosta, Pseudopecten aequivalvis, Tutcheria cingulata, Hastites umbilicatus, Pentacrinus sp.* and *Hybodus reticulatis*,' together with forms indicating the Middle Lias such as *Homoeorhynchia acuta* and *Amaltheus margaritatus*.

Ammonites collected from Ballintoy and Portrush and now housed in the Ulster Museum and the Geology Department, Queen's University, Belfast (prefixed K and Q.U.B. respectively) were examined by Professor D. T. Donovan. None were very accurately localised nor is it known whether the specimens were collected in situ; nevertheless it is significant that only one of the ammonites was from a horizon below the *raricostatum* Zone. The anomalous record is a specimen of *Psiloceras (Caloceras) intermedium* (Portlock) [Q.U.B. 3458], reputedly from Ballintoy

Table 2 Macrofauna of the Sinemurian Lias, White Park Bay

	Locality (see Figure 5)						
	1	2	3	4	5	6	7
Pentacrinoid columnals	–	–	–	x	–	–	–
Echinoid fragments	–	–	–	x	–	x	x
Rimirhynchia sp.	–	–	–	–	–	x	–
Cincta sp.	–	–	–	–	–	–	x
Rhynchonelloid and terebratuloid fragments	–	x	–	x	x	–	x
Astarte sp.	–	–	–	–	x	x	–
Camptonectes mundus Melville	x	x	–	–	x	–	–
C. sp.	x	x	–	x	x	x	–
Cardinia attenuata (Stutchbury)	–	–	–	–	–	–	x
Cercomya sp.	–	–	–	–	x	–	–
Chlamys rollei? (Stoliczka)	–	–	–	–	x	–	–
C. subulatus (Münster)	x	–	–	–	–	–	–
C. textoria (Schlotheim)	–	x	–	–	–	–	–
C. sp.	–	x	–	x	–	–	–
Cucullaea sp.	–	–	–	x	–	–	–
Dacryomya minor (Simpson)	–	–	–	–	x	–	–
Entolium liasianum (Nyst)	–	x	–	–	x	–	–
Goniomya hybrida (Münster)	–	cf.	–	–	x	–	–
G. sp.	?	x	x	?	–	?	–
Grammatodon insons Melville	x	–	–	x	x	–	–
G. sp.	x	x	x	x	x	x	–
Gryphaea arcuata Lamarck	–	?	–	x	–	x	–
G. sp.	–	–	–	x	–	–	–
Homomya sp.	–	–	–	–	?	–	–
Liostrea cf. *hisingeri* (Nilsson)	–	x	–	–	–	–	–
L. sp.	–	x	–	–	–	–	–
Lucina sp.	x	x	–	–	?	–	–
Meleagrinella substriata (Münster)	–	?	–	–	–	–	–
M. sp.	–	–	–	x	–	–	–
Modiolus sp.	x	–	–	–	–	x	–
Nucula sp.	–	–	–	x	–	–	–
Nuculana (Rollieria) bronni (Andler)	–	x	–	–	–	–	–
Oxytoma inequivalve (J. Sowerby)	x	x	–	–	–	x	x
Palaeoneilo galatea (d'Orbigny)	x	x	–	x	x	–	–
P. sp.	–	x	–	–	–	–	–
Placunopsis sp.	–	–	–	x	–	–	–
Plagiostoma sp.	–	–	–	?	–	–	–
Pleuromya sp.	–	x	–	x	x	x	–
Plicatula sp.	x	x	–	–	x	–	–
Protocardia truncata (J. de C. Sowerby)	cf.	x	–	x	x	–	–
P. sp.	x	x	–	–	–	x	–
Pseudolimea acuticostata (Münster)	–	x	–	x	x	–	–
P. sp.	–	–	x	x	x	x	–
Pseudopecten sp.	–	?	x	–	–	–	–
Pteria sp.	–	–	–	x	–	–	–
Pteromya sp.	–	x	–	–	?	–	–
Ryderia doris (d'Orbigny)	–	–	–	x	x	x	?sp.
Tutcheria sp.	x	x	–	–	–	–	–
Unicardium arenacea (Terquem)	?	–	–	–	–	–	–
U. subglobosa (Tate)	–	?	–	–	–	–	–
U. sp.	x	x	–	x	–	–	–
Cheltonia sp.	–	–	–	x	–	–	–
Crucilobiceras sp.	x	–	–	–	–	–	–
Eoderoceras armatum (J. Sowerby)	–	–	x	–	–	–	–
E. sp.	–	–	x	–	–	–	–
Gemmellaroceras tubellum (Simpson)	x	x	x	–	x	–	–
G. sp.	–	x	–	–	–	–	–
'*Hemimicroceras*' cf. *vitreum* (Simpson)	–	–	x	–	–	–	–
'*H.*' cf. *cingendum* Trueman & Williams	–	–	x	–	–	–	–
'*H.*' *sp.*	–	–	x	–	–	–	–
Leptechioceras sp.	x	–	–	–	–	–	–
Oxynoticeras sp.	–	–	–	–	x	–	–
Paltechioceras boehmi (Hug)	x	–	–	–	–	–	–
P. sp.	x	–	–	–	–	–	–
Nannobelus acutus (Miller)	x	x	–	–	–	–	–
indet. belemnite	–	–	–	–	–	x	–
Acteonina sp.	–	–	x	–	–	–	–
Cylindrobullina sp.	–	–	x	–	–	–	–
Procerithium sp.	–	x	–	–	–	–	–
Ptychomphalus sp.	–	?	–	–	–	–	–
Zygopleura berthaudi (Dumortier)	–	–	x	–	–	–	–
Ostracods	x	x	–	–	–	–	–
Fish fragments	x	–	–	–	–	–	–
Wood fragments	x	x	–	–	x	–	–

Grid References of localities 1–7:
1 D 0155 4378; 2 D 0280 4426; 3 D 0230 4426; 4 D 0261 4446;
5 D 0288 4484; 6 D 0288 4481; 7 D 0287 4485.

Table 3 Chronostratigraphic divisions identified in the Lower Jurassic of the Port More Borehole

Substage	Zone	Subzone	Limits assigned
			metres
Lower Pliensbachian	*Tragophylloceras ibex*	*Beaniceras luridum*	(*not present*)
		Acanthopleuroceras valdani	168.35+ to 173.96
		Tropidoceras masseanum	173.96 to 180.82
		Fault zone	180.82 to 180.89?
	Uptonia jamesoni	*Uptonia jamesoni*	?180.89 to ~205.2
		Subzones undivided	
		(see text, p. 82)	~205.2 to ~257.6
		Phricodoceras taylori	~257.6 to ~284
Upper Sinemurian	*Echioceras raricostatum*	*Paltechioceras aplanatum*	~284 to 316.5
		Leptechioceras macdonnelli	316.5 to 330.4
		Echioceras raricostatoides	330.4 to 340.89
		Crucilobiceras densinodulum	(*not proved*)
		Igneous intrusions	337.57 to 337.72
		and	340.89 to 361.29
	Oxynoticeras oxynotum	No evidence for the presence	
	Asteroceras obtusum	of these zones	
Lower Sinemurian	*Caenisites turneri*	Subzones not proved	361.29 to 372.9
	Arnioceras semicostatum	*Euagassiceras sauzeanum*	272.9 to ~414.9
		Agassiceras scipionianum	~414.9 to 437.8
		Coroniceras reynesi	(*not present*)
		Igneous intrusions with hornfels xenoliths	437.8 to 660.7

but in a matrix and preservation closely comparable with the Hettangian of the east coast of Antrim.

From Portrush two significant ammonites were identifiable. These were *Leptechioceras* aff. *macdonnelli* (Portlock) [Q.U.B. 21257] and *Paltechioceras sp.* [group of *studeri* and *favrei* (Hug)] [Q.U.B. 19326]. Both are from the *raricostatum* Zone, *macdonnelli* Subzone, though the latter can also range up into the *aplanatum* Subzone.

From the White Park Bay–Ballintoy area the Ulster Museum collection contained two specimens from the Middle Lias (*spinatum* Zone), probably collected from the drift. These were *Pleuroceras salebrosum* (Hyatt) [K 1253] and *Pleuroceras sp.* [K 1290–1296].

The remaining specimens were either from the *raricostatum* or the *jamesoni* zones and include:

raricostatum Zone: *Crucilobiceras* aff. *crucilobatum* S. S. Buckman [K 2776], *Eoderoceras armatum* (J. Sowerby) [K 1275], *Gemmellaroceras emersoni* (Hoffmann) [K 1272–1274], *Gemmellaroceras tubellum* (Simpson) [K 1243], '*Hemimicroceras*' sp. [K 1249], *Leptechioceras meigeni* (Hug) [K 1282], *Leptechioceras sp.* [Q.U.B. 16307, smaller specimen], *Neomicroceras commune* Donovan [K 1240–1242], *Paltechioceras boehmi* (Hug) [K 2777], *Paltechioceras sp.* [K 1271, K 2778, Q.U.B. 16307]

jamesoni Zone: *Apoderoceras aculeatum* (Simpson) [K 1345], *Apoderoceras hamiltoni* (Simpson) [K 1340, Q.U.B. 17214], *Apoderoceras sp.* [juv.] [K 1284–1289], *Radstockiceras buvignieri* (d'Orbigny) [K 2779, K 1280] (Figured in Wright, 1883, pl. 76, figs. 1, 2).

Two specimens collected from the drift in excavations for a dam for Port Ballintrae waterworks were identified as *Paltechioceras sp.* (cf. *oosteri* Dumortier) or *Leptechioceras sp.* [*raricostatum* Zone] and *Platypleuroceras brevispina* (J. de C. Sowerby) [*jamesoni* Zone].

STRATIGRAPHICAL PALAEONTOLOGY OF THE LOWER JURASSIC OF THE PORT MORE BOREHOLE

A summary of the major features of the Lower Jurassic of this borehole is given in Part I, Chapter 3, and an abbreviated log in Table 3 (above). The main features of the distribution of the ammonites, prepared from identifications made by Professor D. T. Donovan, are shown on Figure 16.

The borehole penetrated the base of the Cretaceous at 168.35 m and entered a thick sequence of dark-grey fossiliferous mudstones with two major groups of intrusive rocks. The fossils were however remarkably unaffected by the heat from these intrusions and a detailed sequence of ammonites has been established in the higher beds:

Tragophylloceras ibex ZONE   168.35 m to ~180.82 m

Acanthopleuroceras valdani SUBZONE   ?168.35 m to 173.96 m

The highest ammonites found, at 168.4 m, belong to the genus *Aegoceras* (*Beaniceras*) but were not specifically identifiable. *Acanthopleuroceras*, indicating the *valdani* Subzone, was found between 168.86 and 173.96 m. Within this range *A. valdani* (d'Orbigny) itself occurs between 169.57 and 170.03 m and *A. maugenesti* (d'Orbigny) between 170.5 and 173.89 m. The zonal index, *Tragophylloceras ibex* (Quenstedt) occurs at 169.44 m and *Liparoceras* at 169.5 m. The base of the subzone is taken at the first appearance of *Acanthopleuroceras* at 173.96 m. The beds above its range and below the Cretaceous are included in the subzone.

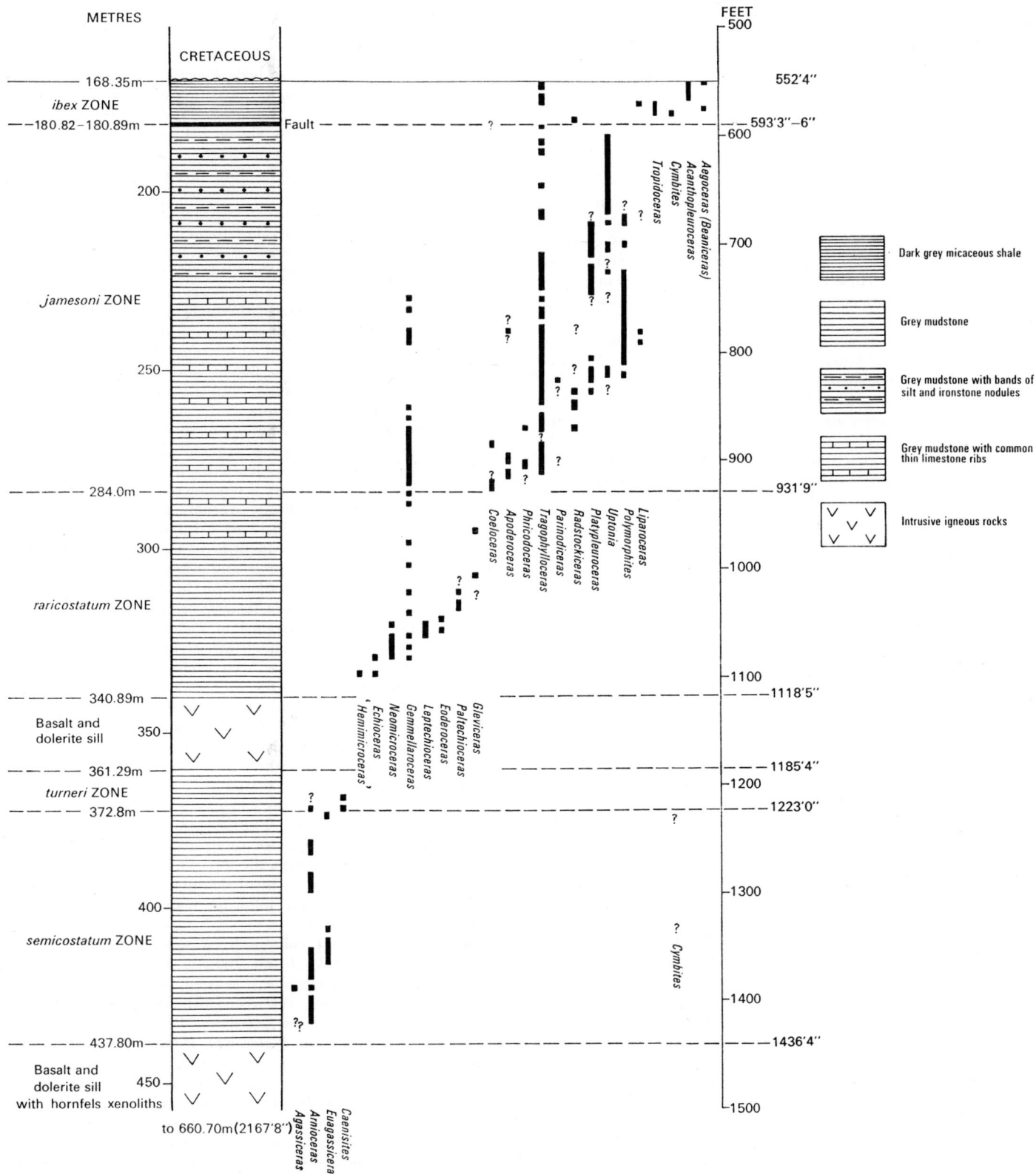

Figure 16 Ammonite ranges from the Jurassic of the Port More Borehole

82 CHAPTER 14 JURASSIC ROCKS

Tropidoceras masseanum SUBZONE   173.96 m to ~180.82 m

The lowest occurence of *Tropidoceras* at 179.37 m marks the lowest evidence for this subzone and for the *ibex* Zone. Between 180.82 and 180.9 m sheared mudstones indicate a fault and the 1.5 m of beds immediately above contain bivalves and an indeterminate aegoceratid so that their horizon cannot be accurately determined. They are provisionally included in the base of the *ibex* Zone. *Tropidoceras ellipticum* (J. Sowerby) occurs at 178.9 m and *T.* aff. *stahli* Spath at 175.46 m. *Tropidoceras* was also found to range up into the *valdani* Subzone to 173.84 m. Associated ammonites include *Cymbites* (*Metacymbites*) *sp.* between 176.83 and 177.95 m and *Radstockiceras sp.* at 178.92 m. The specimens of *Aegoceras* (*Beaniceras*) cf. *luridum* (Simpson) around 175.3 m and *Tragophylloceras carinatum* Howarth & Donovan at 174.57 m are at notably low horizons compared with the ranges of these taxa known in England.

Apart from the ammonites a varied bivalve and brachiopod fauna is present. The latter includes some specimens of *Discinisca* and also rhynchonelloids and terebratuloids. Belemnites are also common. The bivalves include *Laevitrigonia*, which is a form characteristic of the *jamesoni* and *ibex* zones in central England (Melville, 1956). Lithologically the sediments of this zone can be described as rather hard slightly calcareous mudstones, often finely micaceous and with thin bands of incipient ironstone nodules. A thin siltstone occurs at 173.1 m.

Uptonia jamesoni ZONE   ?180.89 m to ~284.0 m

Below the highly sheared material of the fault between ~180.82 and 180.89 m a very thick sequence of finely micaceous mudstones with occasional ironstone nodules and thin siltstones can be divided between the *jamesoni* and *raricostatum* zones and indicates a *jamesoni* Zone of over 100 metres in thickness. The generally abundant ammonite fauna of this sequence has cast new light on the ranges and occurrence of the Polymorphitidae and suggests lines for a revision of the subzonation. The ranges of the genera of ammonites are shown in Figure 16.

Consideration of the detailed ranges of species established from these specimens indicates that the four-fold subdivision of the *jamesoni* Zone proposed by Dean, Donovan and Howarth (1961) cannot readily be applied to this sequence, and it is currently being studied in comparison with the details of the thick *jamesoni* Zone sequence in the Llanbedr (Mochras Farm) Borehole (Ivimey-Cook, 1971). It is possible that a three-fold subzonation based on a *brevispina* Subzone containing the main range of both *Polymorphites* and *Platypleuroceras* may prove to have value. Pending the completion of this examination no subzonal classification is suggested for the middle part of the Zone.

Immediately below the fault at 180.89 m, a specimen of *Tragophylloceras loscombi* (J. Sowerby) occurred at 181.28 m. This form is usually associated with a horizon high in the *ibex* Zone and it must be a possibility that these beds are in a fault-bounded slice associated with the movement of the main fault plane, but no lithological evidence for this is available. The *Coeloceras?* at 181.76 m is compatible with a *jamesoni* Zone age whilst the highest *Uptonia* was found at 183.03 m.

The highest part of the sequence, probably equivalent to the *jamesoni* Subzone, between ?180.89 and 205.2 m is dominated by *Uptonia*. *Uptonia bronni* (Roemer) occurs in the lowest 6 m between 204.3 and 198.4 m and overlaps the lowest *U. jamesoni* (J. de C. Sowerby) at 201.17 m, the latter species is predominant between 196.7 and 185.1 m. This order of occurrence contrasts with that known elsewhere (Dean, Donovan and Howarth, 1961, p. 464) where *U. bronni* occurs in the upper part of the range of *Uptonia*.

The middle part of the *jamesoni* Zone between about 205.0 and 257.6 m contains the ranges of *Polymorphites* and *Platypleuroceras* which are almost coincident in this borehole but the detailed pattern of occurrences shows variations some of which may prove to be significant.

In the lowest part of these beds *Platypleuroceras* is found from 257.6 m up to about 246.2 m, so co-existing with the earliest group of *Uptonia* (255.7 to 248.6 m). The earliest *Polymorphites* occur in the upper part of this thickness above 251.3 m at which horizon *Liparoceras* (*Parinodiceras*) also occurred.

Between 246.5 m and 222.0 m four species of *Polymorphites* can be identified; *P. costatus* (Quenstedt) and *P. lineatus* (Quenstedt) are more common in the lower beds than *P. mixtus* (Quenstedt) and *P. quadratus* (Quenstedt). The highest *Gemmellaroceras* occurs at 230.0 m and an *Apoderoceras?* at 234.3 m. Between 246.2 and 231 m, however, neither *Platypleuroceras* nor *Uptonia* was recorded. Above 246.2 m *Platypleuroceras* reappears in some numbers and then gradually becomes scarce; the highest specimen was found at 205 m. Above 222 m *Polymorphites* is also rare though it does persist slightly longer to 202 m. At the same time *Uptonia* becomes rather more common up to 205 m and thereafter forms the majority of the ammonite fauna. Numerous small *Tragophylloceras* occur almost throughout the zone.

Beds approximating to the *taylori* Subzone occur below 257.6 m and the base of the subzone is provisionally placed below the lowest *Coeloceras* cf. *pettos* (Quenstedt) at 283.7 m. The exact horizon for the zonal boundary is not however clear. The range of *Phricodoceras* (280.14 to 266 m) overlaps part of the range of *Liparoceras* (*Parinodiceras*) (276.12 to 251.3 m), although the subzonal index species was not identified. These beds contain the main abundance of *Apoderoceras* and *Gemmellaroceras*. The only ammonites other than *Gemmellaroceras* found for many metres below this were specimens of *Gleviceras* occurring at 292.9 m and 307.2 m. The highest occurrence of *Paltechioceras*, indicating a horizon in the *raricostatum* Zone, is at 307.4 m. The 23 m of strata between 284 and 307 m have been excluded from the *jamesoni* Zone as the highest beds of the *raricostatum* Zone are known to contain only a limited ammonite fauna (Getty, 1973).

Lithologically the *jamesoni* Zone is dominated by dark-grey mudstone that is intermittently shaly and silty and contains some pyrites. The higher parts of the zone have nodules of calcareous ironstone and four widely scattered 5–8-cm siltstone bands. Below 252 m the beds are more calcareous and contain wisps and thin partings of limestone. The sequence contains numerous listric surfaces and evidence of small-scale faulting and minor shearing, particularly towards the top. The mudstones contain a diverse and locally abundant bivalve fauna and also brachiopods, crinoids and belemnites. The following taxa have been identified by Mr R. Clark:

Serpulid tubes, crinoid columnals, *Rimirhynchia*, *Spiriferina* (at 186.5 and 270 m) *Cincta* (below 260 m), *Arcomya*, *Astarte*, *Bakevellia?*, *Camptonectes*, *Eotrapezium*, *Goniomya*, *Grammatodon*, *Gryphaea*, *Hippopodium*, *Laevitrigonia* (between 181 and 256 m), *Liostrea*, *Modiolus*, *Nuculana* (*Rollieria*), *Oxytoma*, *Palaeonucula*, *Palaeoneilo*, *Parainoceramus*, *Parallelodon*, *Pholadomya*, *Pinna*, *Placunopsis*, *Plagiostoma*, *Pleuromya*, *Protocardia*, *Pseudolimea*, *Pseudopecten*, *Ryderia*, *Tutcheria*, *Unicardium*, various small gastropods, *Dentalium* (*Antalis*) and belemnites. An ammonite aptychus was found at 202.8 m. Ammonites were notably dominant in the macrofauna between 213 and 236 m where they are much more common than the bivalve remains. The list of genera is closely comparable with that recorded from the *jamesoni* Zone in the more condensed English sequences.

Echioceras raricostatum ZONE   ~284 m to 340.89 m

The scarcity of ammonites in these beds greatly handicaps attempts to establish the subzones present. However the following divisions may be postulated:

SUBZONE	DEPTH
Paltechioceras aplanatum	~284–316.5 m
Leptechioceras macdonnelli	316.5–330.4 m
Echioceras raricostatoides	~330.4–?340.89 m (on igneous intrusion)
Crucilobiceras densinodulum	not proved

No evidence was found for the basal *densinodulum* Subzone. The lowest ammonites found—at 335 m—were '*Hemimicroceras*' and *Echioceras raricostatum* which indicate the *raricostatoides* Subzone, and this can be taken up to the lowest occurrence of *Neomicroceras* at 330.4 m. One *Echioceras sp.* occurs just above this level at 329.1 m.

The range of *Leptechioceras*, 321.5 to 324.3 m, is an indicator of the minimum extent of the *macdonnelli* Subzone. *Neomicroceras* is frequently associated with it in the subzone and is here used to give the subzone a range from 316.5 to 330.4 m (see Donovan, 1966). These beds also contain rare *Eoderoceras* and the lowest *Gemmellaroceras* which occur down to 329.9 m.

The range of *Paltechioceras* usually commences low in the *macdonnelli* Subzone, but in this borehole it was identified only above the range of both *Leptechioceras* and *Neomicroceras* and these beds have been arbitrarily placed in the *aplanatum* Subzone. This subzone has been continued above the highest *Paltechioceras* at 307.4 m through the beds with rare *Gleviceras* and *Gemmellaroceras* up to the base taken for the *jamesoni* Zone at 284 m.

Lithologically the *raricostatum* Zone is similar to the upper part of the *jamesoni* Zone. The dark-grey mudstones are frequently very silty and also often slickensided with listric surfaces and some calcite-filled slip surfaces. The sequence also contains scattered ironstone nodules and occasional partings and wisps of thin limestone. Siltstone beds are more common including a 1.22-m bed below 300.5 m and thinner beds below to 326 m. A thin basaltic intrusion occurred between 337.56 and 337.72 m, just above the major complex sill whose top is at 340.89 m.

The non-ammonite fauna of these beds is notably less varied and less abundant than that of the *jamesoni* Zone. Mr Clark has recognised the following taxa: serpulid tubes, crinoid columnals, rhynchonelloids, bivalves including *Anningella*, *Arcomya*, *Camptonectes*, *Chlamys*, *Goniomya*, *Grammatodon*, *Gryphaea*, *Hippopodium*, *Liostrea*, *Myoconcha*, *Nuculana* (*Rollieria*), *Oxytoma*, *Palaeoneilo*, *Parainoceramus*, *Pholadomya*, *Plicatula*, *Protocardia*, *Pseudolimea*, *Pseudopecten*, *Pteromya?*, *Ryderia* and *Unicardium*. Rare gastropods were present and also belemnites and fish fragments.

This fauna is consistent with the fauna found at White Park Bay in this zone but most of the genera are quite long-ranging and little detailed stratigraphic significance has been established.

The basalts and dolerites comprising the sill continued between 340.89 and 361.29 m and the heat of intrusion had hornfelsed the underlying mudstone for over 1 m below the contact.

No evidence was found for the presence of the *oxynotum* or *obtusum* zones or for the upper part of the *turneri* Zone.

Caenisites turneri Zone ?361.3 m+ to 372.9 m

The presence of *Caenisites* at 369.8 and 372.82 m indicates the *turneri* Zone; other ammonites present in these beds include *Cymbites?* at 371.9 and 372.85 m and *Arnioceras?* at 369.3 m. These are not adequate to establish a subzonal identity for these beds, although the continuous nature of the sedimentation would suggest that they represent the lower beds of this zone.

A very sparse non-ammonite fauna is present in these beds including *Camptonectes*, *Chlamys*, *Palaeoneilo galatea* (d'Orbigny), *Tancredia?* and *Dentalium*. Lithologically the sequence contains grey mudstones, then micaceous siltstones and thin argillaceous mudstones passing into calcareous mudstones.

Arnioceras semicostatum Zone 372.9 m to 437.8 m

All the lower beds above the main intrusive suite starting at 437.8 m are placed in the *semicostatum* Zone. Their ammonite fauna is sparse apart from the occurrence of *Arnioceras* down to 431.75 m. The occurrence however of *Euagassiceras?* at 399.1 m, *E. resupinatum* (Simpson) at 408 to 409 m, and *E. sp.* between 405.1 and 414.9 m suggests the *sauzeanum* Subzone within this range. The range of this subzone has been extended upwards through beds with *Arnioceras semicostatum* (Young & Bird) and *A. miserabile* (Quenstedt) up to the horizon of the lowest *Caenisites* which positively indicates the *turneri* Zone.

In this borehole there is a small gap between the records of *Euagassiceras* and the highest *Agassiceras* which is taken as indicative of the *scipionianum* Subzone. *Agassiceras?* was found between 422.5 and 433.7 m, *Cymbites?* at 422.2 m and *Arnioceras bodleyi* (J. Buckman) at 430.4 m. No further ammonite evidence was found in the few metres of beds above 437.8 m to suggest the presence of the *reynesi* Subzone and it is concluded that this is not present.

Lithologically these beds continue the dark-grey shaly and silty mudstones found throughout the Lower Lias in this borehole and also contain thin wispy calcareous bands and scattered ironstone nodules and thin beds. Between 417 and 419 m the mudstone is notably silty with some fine sand and sandy layers. These more arenaceous layers are again seen, accompanied by *Gryphaea arcuata* Lamarck, between 431 and 436.5 m but the beds do not get progressively sandier downwards. The lowest beds seen are hornfelsed mudstones above the igneous contact. These beds do not contain the sequence of pale-brown calcareous sandstones which in north Londonderry has yielded *Arnioceras* and *Euagassiceras* of the *sauzeanum* Subzone, though the presence of *Lingula* at 412.9 m compared with its presence in the *semicostatum* Zone of the Magilligan townland. The bivalve fauna includes *Anningella*, *Camptonectes*, *Chlamys*, *Gryphaea arcuata* (particularly common below 424 m), *Modiolus*, *Nuculana* (*Rollieria*) *bronni?* (Andler), *Oxytoma*, *Palaeoneilo galatea*, *P. oviformis* Troedsson, *Plagiostoma giganteum* J. Sowerby (below 427 m), *Protocardia* and *Pseudopecten*. *Dentalium sp.*, indeterminate belemnites and fish debris also occur.

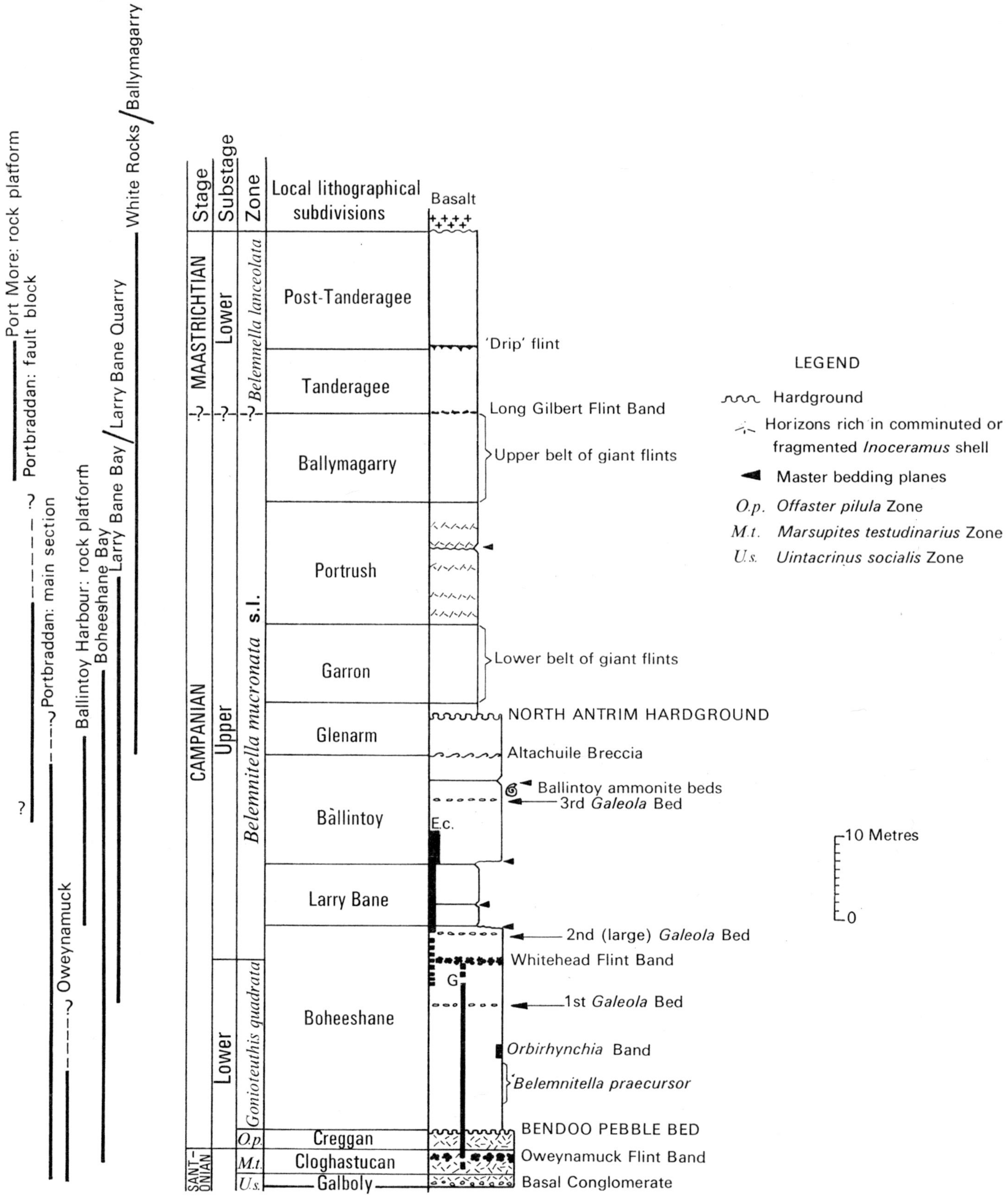

84 CHAPTER 15 CRETACEOUS ROCKS

Figure 17 Summarised stratigraphy of the White Limestone with positions of marker horizons. The vertical columns marked *E. c.* and *G.* show the ranges of *Echinocorys* ex gr. *conica* and *Gonioteuthis* respectively

CHAPTER 15
Cretaceous rocks

The Cretaceous rocks of Northern Ireland are all of Late Cretaceous age. Two formations are recognised—the Hibernian Greensands, ranging from Cenomanian to Santonian, and the Chalk, or Ulster White Limestone, ranging from Santonian to Maastrichtian. In the map-area true greensands are not developed and the base of the Chalk is marked by a conglomerate of Santonian age.

The Chalk or Ulster White Limestone Formation of Sheet 7 outcrops in three main areas: an eastern area stretching from the Port More Fault near the eastern sheet boundary to Port Braddan, where the outcrop is cut off by the Port Braddan Fault; a central, almost totally Drift-covered area to the north of Bushmills, where the Chalk reappears on the south side of the Port Braddan Fault; and a western area stretching from just west of Dunluce Castle to Portrush, where the outcrop is terminated by the Portrush Fault. Throughout the sheet the Chalk rests with major disconformity on Lower Lias, and is overlain unconformably by the Lower Basalts and/or Tuff of the Tertiary Antrim Lava Series. The main outcrop forms a coastal strip interrupted by downfaulted blocks of Tertiary basalt. Within this coastal strip the Chalk is exposed in vertical and generally inaccessible cliff sections (Plate 25), except where the Lias contact is at or about sea-level, in which case landslipping modifies the coastal morphology to a greater or lesser extent. The contrast between these two types of coast can be observed in Boheeshane Bay, north of Ballintoy (see Plate 35), where a small area of landslip fronts the cliffs in the central part of the bay; to the east (left) of the fault seen at the edge of the landslipped area the Lias is well below sea-level, and the cliffs are correspondingly sheer. The landslip effect is even more marked in White Park Bay, where the coastline transects an elongate dome cored by Lias, and the cliffs have receded some 0.5 km from the sea; on the seaward limb of this structure large collapsed chalk stacks may be seen during periods of very low tide.

The thickest and most complete Chalk successions in Northern Ireland accumulated in two major depositional basins—the North Antrim Basin and the East Antrim Basin—lying to the north-west and south-east of the Dalradian 'Highland Border Ridge' respectively. By taking the north and east coast successions in conjunction it has been possible to build up a composite standard succession for these areas, much of the data for which has been derived from the present sheet. Towards the margins of the depositional basins condensation and the presence of non-sequences result in modifications of the standard succession, but such modifications are of only minor importance here. The Chalk of Sheet 7, together with that of the adjacent Ballycastle (8) Sheet to the east, belongs to the North Antrim Basin, which is itself part of the larger Hebridean Mesozoic Basin.

The Sheet 7 succession is of the order of 113 m[1], and covers a stratigraphical range from the Santonian (*Uintacrinus socialis* Zone) to the lower part of the Lower Maastrichtian (*Belemnella lanceolata* Zone). Higher horizons, represented by the *B. occidentalis* Zone recognised at Ballycastle, may be present in the (inaccessible) uppermost beds east of Portrush, but it has not been possible to establish this point. The stratigraphical framework used here is taken from the standard established by Fletcher (*in press*). Of Fletcher's fourteen members comprising the Ulster White Limestone Formation, six have their type sections within the area of the sheet.

STRATIGRAPHY

Two thin units, the upper approximately twice the thickness of the lower, defined by three well-marked erosional separation-planes, are a characteristic feature of many of the coast sections of Sheet 7. This double unit, which is known as the Larry Bane Chalk Member from its type section at Larry Bane Head, is of particular importance for two reasons:

1 It provides a stratigraphical marker by means of which the succession can be subdivided into pre-Larry Bane (30.65 m), Larry Bane (7.28 m) and post-Larry Bane (74.29 m) sequences respectively, the thicknesses referring to Sheet 7 only.

2 The base of the lower unit (that is Larry Bane A) occurs 3.66 m above the level of extinction of the belemnite genus *Gonioteuthis*, and thus approximately marks the base of the *Belemnitella mucronata* Zone (s.l.) and of the Upper Campanian Substage. Larry Bane A is indicated in the photographs of coast sections (Plates 28 and 33) for ease of reference.

Within the pre-Larry Bane Chalk a marked lithological change takes place at an erosion surface with associated glauconitised chalk pebbles 6.4 m above the base of the succession. This marker is named the Bendoo Pebble Bed and it separates a tripartite Galboly–Cloghastucan–Creggan sequence of chalk relatively rich in comminuted *Inoceramus* shell from essentially *Inoceramus*-free chalk above. Wolfe (1968) introduced the term '*Inoceramus* Chalk' for these bioclastic chalks. The erosion surface is comparatively weakly developed in the basinal succession of Sheet 7, but towards the margins passes laterally into a stromatolite-capped hardground complex, for example the Mulatto of the Belfast area.

The post-Larry Bane succession can similarly be divided at the North Antrim Hardground complex[2] *within* the Glenarm Member, for which the reference section is the Larry Bane Quarry (p. 98). As noted previously (Wood *in* Wilson, 1972), the North Antrim Hardground is so called to distinguish it from a similar group of hardgrounds—collectively known as the South Antrim Hardgrounds—which is developed in the Belfast area at a higher horizon (Portrush Chalk), but which has been confused with it. Although there is no obvious change in sediment type at the North Antrim Hardground, a change is apparent in the resistivity profile at this level in the Port More Borehole. There is also a change in flint type. Below the hardground courses of burrow-fill flints are the rule, with irregular thin tabular flints developing in the thinner-bedded units. The succession above the hardground is characterised by two belts of giant flints (including ring flints and locally true paramoudras) separated by a unit of thin-bedded chalks with more or less continuous tabular flints and several horizons enriched in large fragments of *Inoceramus* shell. The North

---

1 Wolfe (1968, p. 270) reported up to 31 per cent shortening of macrofossils in the White Limestone as a result of compaction prior to final lithification of the original soft chalk sediment. Direct comparison of sediment thicknesses with those of correlative successions in other area may therefore be misleading unless the compaction factor is taken into account.

2 It should be noted that in the definitive account of the lithostratigraphy of the Ulster White Limestone (Fletcher, *in press*) this complex is termed the North Antrim Hardgrounds, and accordingly the plural form should be understood throughout this memoir.

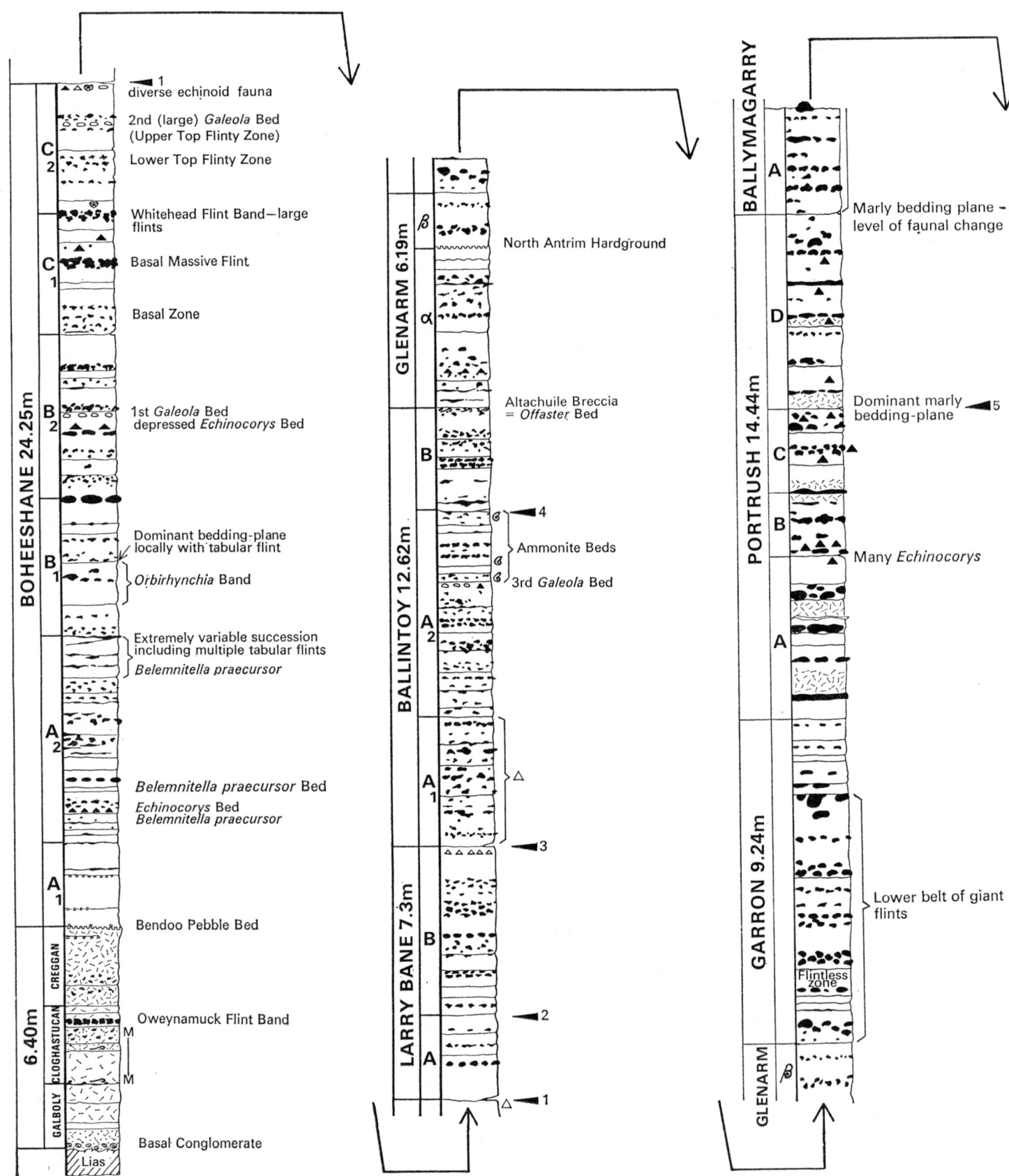

Figure 18 Composite lithostratigraphic succession of the White Limestone

STRATIGRAPHY 87

Antrim Hardground occupies a position almost exactly half way up the Sheet 7 succession, 55 m above the base.

A further important lithological marker is provided by the Long Gilbert Flint Band, a conspicuous semi-continuous thick tabulate flint developed in the Portrush area 36 m above the North Antrim Hardground, and some 21 m below the local top of the succession. This marker, which can be seen in the southern wall of the central Ballymagarry Quarry east of Portrush, terminates the higher belt of giant flints and marks the entry of a fauna of Maastrichtian aspect but without the diagnostic belemnite genus *Belemnella*. The level of the Long Gilbert Flint is tentatively taken as the Campanian–Maastrichtian boundary (see discussion, pp. 109–110). The overlying beds see a return to the normal burrow-fill type of flint, although in the uppermost part of the succession giant flints are again developed.

A summarised version of the stratigraphy with the position of most of the major lithological and/or palaeontological marker horizons indicated is given in Figure 17. It will be seen that the greater part of the Causeway Coast succession is Campanian, of which approximately three-quarters belongs to the Upper Campanian Substage. The detailed stratigraphy is presented as two double columns (Figure 18) covering the pre- and post-North Antrim Hardground successions respectively. The original measurements (by Fletcher) were made in Imperial units and are accurate to the nearest inch up to the top of the Ballymagarry Chalk. The post-Ballymagarry succession, however, is accessible only in a poorly exposed and complex faulted area of the Ballymagarry Quarry, and the measurements are uncertain. This applies in particular to the succession above the Tanderagee Chalk, which has been tentatively correlated with the lower part of the Ballycastle Maastrichtian, that is the Port Calliagh Member of Fletcher's scheme (see p. 111). These highest beds have been designated numerically for convenience.

The stratigraphical classification for Sheet 7 is as follows:

Post-Larry Bane sequence	post-Tanderagee succession (numbered units)
	Tanderagee Chalk (A–E)
	Ballymagarry Chalk (A–C)
	Portrush Chalk (A–D)
	Garron Chalk (undivided)
	Glenarm Chalk β
	——— *North Antrim Hardground* ———
	Glenarm Chalk α
	Ballintoy Chalk (A–B)
Larry Bane sequence	Larry Bane Chalk (A–B)
Pre-Larry Bane sequence	Boheeshane Chalk (A–C)
	——— *Bendoo Pebble Bed* ———
	Creggan Chalk
	Cloghastucan Chalk
	Galboly Chalk
Hibernian Greensands	Basal Conglomerate
	——— *disconformity* ———
	LIAS

The Santonian part of the Sheet 7 succession is relatively thin and lacking in flint compared with the more flinty equivalents in east Antrim, where the Santonian reaches its maximum thickness. For this flinty development, Fletcher proposed two members (Galboly and Cloghastucan) which, although defined lithostratigraphically, effectively correspond to Chalk with *Uintacrinus* and Chalk with *Marsupites*. This classification is followed in the present sheet although in some localities due either to absence

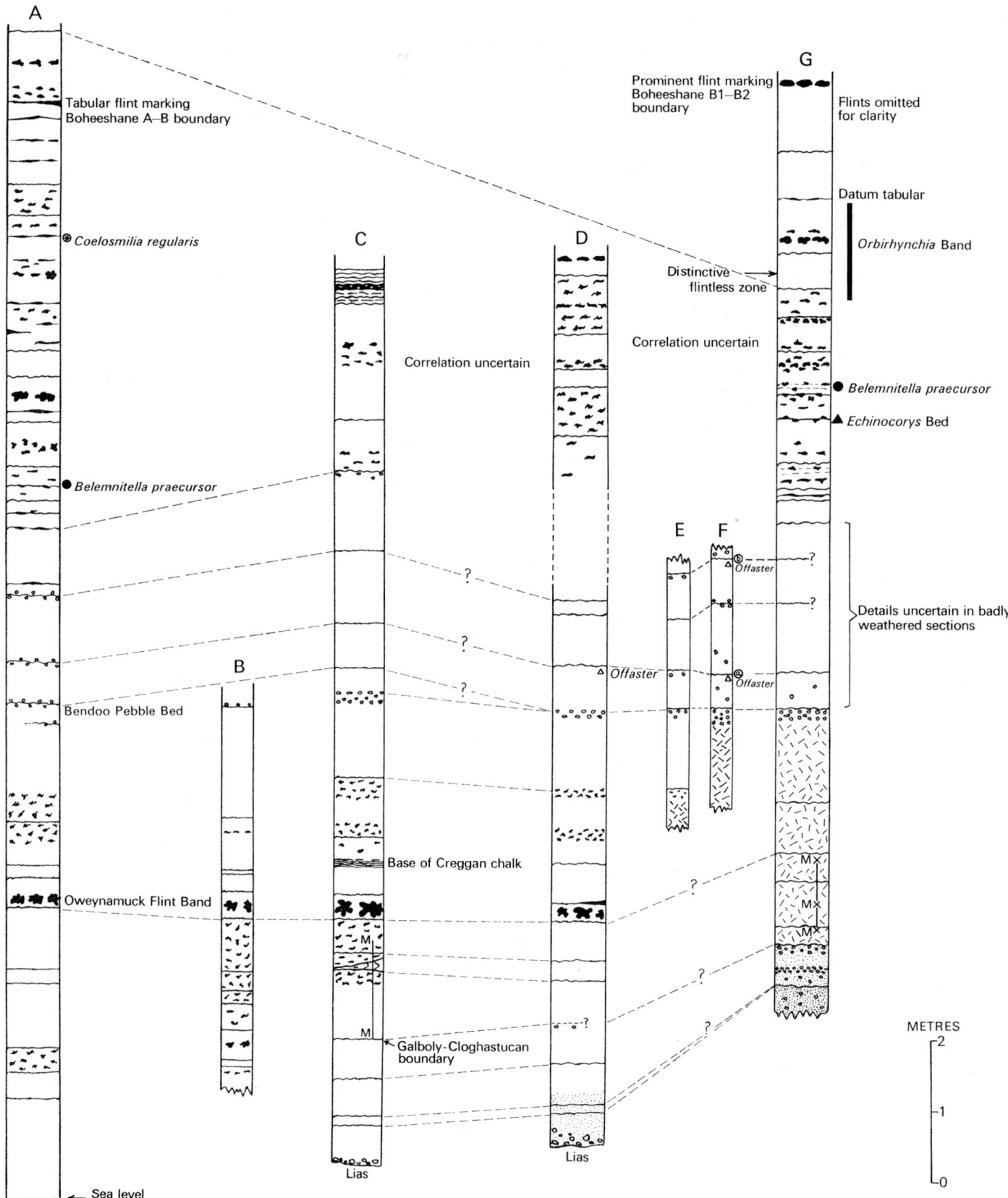

Figure 19 Variation in the Santonian and part of the Lower Campanian succession from Boheeshane (A), through the quarry between Ballintoy Harbour and Dunnaglea (B), Oweynamuck (C), White Park waterfall (D), fallen block exposures in the central (E) and western (F) parts of White Park Bay, to Portbraddan (G).
M denotes the occurrence of *Marsupites*. Based on sections measured by R. E. H. Reid, Queen's University, Belfast

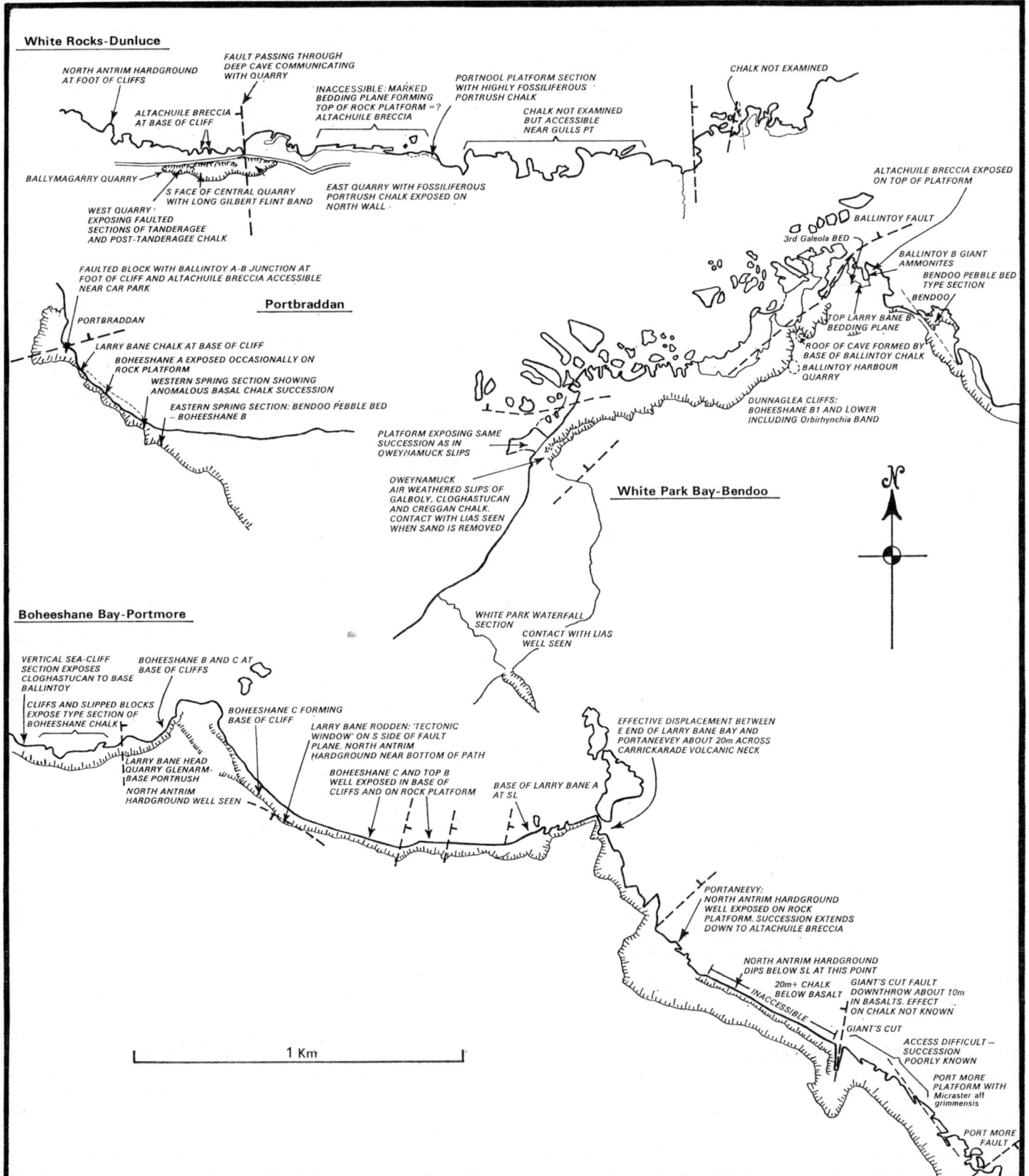

Figure 20 Locations of important sections in the coastal outcrop of the Cretaceous rocks

of flint or to poor palaeontological control it has not been possible to distinguish the Galboly and Cloghastucan members as separate entities. These basal members exhibit marked lateral variation both in thickness and detailed stratigraphy (Figure 19), since it is this part of the succession which occupies the irregularities in the pre-Cretaceous floor; the Creggan Member is considerably less variable, although it tends to become flintless with reduction in thickness, for example at Portbraddan. The local reference section for the bottom two members is taken at Oweynamuck rather than at the thicker succession in Boheeshane Bay because the former is much more accessible and the palaeontology has been studied in detail.

The stratigraphical column (Figure 18) has been put together from the following key sections:

1 post-North Antrim Hardground succession: White Rocks and Ballymagarry Quarry
2 Glenarm α: Larry Bane Head Quarry
3 upper 1/3 Ballintoy A and Ballintoy B: Ballintoy Harbour
4 lower 2/3 Ballintoy A: Larry Bane Head
5 Boheeshane C and Larry Bane: Larry Bane Bay and Larry Bane Head
6 Creggan and Boheeshane A, B: Boheeshane Bay
7 Galboly and Cloghastucan: Oweynamuck

All the pre-North Antrim Hardground reference sections are found in the short coastal stretch between Carrickarade and Ballintoy Harbour (see Figure 20 and Plates 31 and 33), a linear distance of about 2 km. Effectively, however, the entire lower half of the Sheet 7 succession is represented in the Larry Bane Head and Quarry sections, together with the nearby cliff section in the centre of Boheeshane Bay. The Larry Bane Head Quarry also exposes the lower part of the post-North Antrim Hardground succession, but the section is unstable and dusty as the result of blasting. This higher succession is best examined in the western outcrop area, where the sea cliffs known as the White Rocks and the adjoining Ballymagarry Quarry provide a remarkable combined section from the Altachuile Breccia at the top of the Ballintoy Chalk to the highest beds exposed in the Sheet.

The composite stratigraphical column for Sheet 7 derives from two superimposed sections taken from areas some 17 km apart. Evidence from the Port More Borehole and from outcrop suggests that there are lateral variations in thickness of certain units in the post-North Antrim Hardground succession between the two areas, so that the section as given may not be representative of any one area. The sediments of the western area at outcrop are considerably less indurated than those in the east, with the result that fossils are easier to extract. This induration reaches its maximum intensity in Rathlin Island in the adjacent Ballycastle (8) Sheet, notably in Altachuile Bay on the north side of the island.

DETAILED STRATIGRAPHICAL PALAEONTOLOGY

THE HIBERNIAN GREENSANDS FORMATION

On the published map a thin unit classified as Hibernian Greensand is shown at the base of the Cretaceous succession, resting with major unconformity on the Lower Lias. This unit is coded h5a–c, of Lower, Middle and Upper Chalk age, but a note by the explanatory vertical section states that the Hibernian Greensand in the area of Sheet 7 appears to be only of Upper Chalk age. Strictly speaking, however, this deposit is an erosional relict of the uppermost part of a complex of clastic sediments which spans a stratigraphical range from Cenomanian to Santonian.

This bed may be regarded as the basal unit of the Chalk succession and is exposed at two localities in White Park Bay, viz. the base of the slipped masses at Oweynamuck [D 029 448], and the bottom of the waterfall section [D 028 443] in the recessed cliffs forming the main amphitheatre. The Lias is overlain by a concentrate of phosphatised pebbles of various provenances enclosed in a matrix of chalk heavily contaminated with glauconite grains. The concentrate is essentially a basal conglomerate, incorporating material derived from the Lower Lias, together with lightly phosphatised echinoids and bivalves of Upper Cretaceous age. Echinoids collected from Oweynamuck include shape variants of *Echinocorys* indicative of the highest part of the *Micraster coranguinum* Zone in southern England, a zonal position confirmed by the record (Hancock, 1961) of *Conulus albogalerus* Leske. No exactly comparable *Echinocorys* have been found to date elsewhere in Northern Ireland, but it is thought likely that this fauna equates with that of the highest part of the Hibernian Greensands—i.e. the Upper Glauconitic Sandstone of Reid (1971)—or the base of the overlying Sponge Beds. The fauna of the conglomerate also includes *Spondylus spinosus* (J. Sowerby) and fragments of thick-shelled *Inoceramus*, and in this respect compares with the youngest Hibernian Greensands fauna, but other characteristic elements—notably the rhynchonellid and terebratulid assemblages—appear to be totally unrepresented. In view of the fact that only a tiny fraction of the whole complex is preserved in the basal conglomerate it is uncertain how much Hibernian Greensands was deposited in the northern area. It is arguable, moreover, that the fossils derived from the *coranguinum* Zone are enclosed in a matrix of post-*coranguinum* Zone sediment, i.e. of *Uintacrinus socialis* Zone age, which is contaminated with glauconite derived from the erosion of pre-existing Hibernian Greensands members. The possibility of a correlation between this basal unit in the Sheet 7 succession and the Cloghfin Sponge Beds developed in the East Coast area (Island Magee) has not been satisfactorily resolved, although evidence from the *Echinocorys* variants present at Oweynamuck suggests that at least the lowest part of the Sponge Beds is represented.

THE ULSTER WHITE LIMESTONE OR CHALK FORMATION

The pre-North Antrim Hardground succession

The Basal Conglomerate is overlain by a tripartite succession of '*Inoceramus* Chalk' delimited upwards by the Bendoo Pebble Bed, which embraces condensed Santonian deposits and the basal part of the Lower Campanian; in terms of the standard English zonation this corresponds to a range from the *socialis* Zone to the top of the lower subzone (*Echinocorys depressula*) of the *Offaster pilula* Zone. These beds are most conveniently studied in the slipped masses at Oweynamuck (Figure 19, Section C; Plate 26) and in smaller blocks near Dunnaglea [D 033 450], where weathering has accentuated the sedimentary features. They can also be examined in Boheeshane Bay, the quarry to the west of Ballintoy Harbour [D 035 451], the White Park Bay amphitheatre waterfall section and in a poorly exposed and anomalous section near Portbraddan (Figure 19, Sections A, B, D and G). Lateral variation in the lowest two members (Galboly and Cloghastucan) is considerable, even within a few metres, as can be clearly seen in the various sections at Oweynamuck; in consequence detailed correlation with expanded sequences, for example Boheeshane Bay, and also Murlough Bay on the adjoining Sheet (8) presents considerable difficulty. In the coastal stretch between Boheeshane Bay and Portbraddan, the various units of this tripartite sequence exhibit progressive condensation and concomitant loss of flint, so that at Port Braddan itself the entire pre-Boheeshane succession is flintless.

The Galboly and Cloghastucan members are conveniently taken together, since their distinction is based more on palaeontological than on lithological criteria. These units respectively constitute the reduced correlatives of the two crinoid zones (*Uintacrinus socialis* and *Marsupites testudinarius*) of the standard zonation. The sediment is greyish-cream rather than white, and has a distinctly 'gritty' texture resulting from the presence of comminuted *Inoceramus* shell and other bioclastic debris. Irregular and/or wavy bedding is usual, with the bedding surfaces emphasised by thin skins of greyish-green marl, which often include fish scales. Interrupted sedimentation is shown by minor discontinuous erosion surfaces marked by concentrations of rounded glauconitised (green) chalk pebbles; locally these erosion surfaces appear to cut down into one another. Many of the planes etched out in air-weathered sections such as Oweynamuck and Dunnaglea are probably more the result of secondary diagenesis (for example solution/compaction phenomena) than primary sedimentary features, and their value in correlation is limited. The green-pebble erosion surfaces are of more significance as they mark definite sedimentary events, but correlation with similar surfaces in other areas is uncertain at present.

GALBOLY CHALK (Variable thickness: the lowest three members together total 6.7 m at Oweynamuck)

The Basal Conglomerate is succeeded by a bed of markedly glauconitic chalk terminating in a green-pebble erosion surface. The higher part of the sequence is considerably less glauconitic, with the glauconite restricted to scattered grains and to coatings on pebbles and sedimentary surfaces. The glauconite content is almost certainly derived in large part from the reworking of glauconitic sands belonging to the Hibernian Greensands. The top of the unit is taken at the base of a well-marked zone of wavy bedding which marks the base of Cloghastucan Chalk (see below) and which effectively delimits the upward range of *Uintacrinus*. It should be noted that flints are absent from this member in White Park Bay but are found in other localities where the succession is expanded, for example Boheeshane Bay. Where found they are of small size and spindle-shaped, i.e. of silicified (crustacean) burrow-fill type (Bromley, 1967).

The fauna is limited. Dissociated brachials and calyx plates of *Uintacrinus* range up to the top, but have not been collected so far from the glauconitic bed above the basal conglomerate. The hard nature of the matrix is such that these small plates can be collected only where exposed on weathered surfaces; it would be exceedingly difficult to recognise them in a sea-battered section or in a core. With the exception of echinoids and belemnites, the remaining fauna comprises:
elongate lituolid(?) foraminifera
Neoflabellina sp.
Porosphaera globularis (Phillips)
indeterminate (hexactinellid) sponges preserved as green pebbles
Meliceritites dollfussi Pergens
Glomerula gordialis (Schlotheim)
Parasmilia fittoni Milne Edwards & Haime
Coelosmilia?
Terebratulina rowei Kitchin
T. striatula (J. de C. Sowerby *non* Mantell)
teleostean bones and scales
Pseudocorax laevis Leriche; tooth

The echinoids include a small *Micraster*, and flat-topped variants of *Echinocorys striata* of the types that characterise the middle part of the *socialis* Zone in Thanet. Small actinocamaxoid belemnites, provisionally determined as *Actinocamax verus* (Miller), are seen in section, but no complete specimens have been found. *Gonioteuthis*—i.e. early *granulata* forms—have not been recorded, although they occur infrequently at this level in Thanet. Where *Uintacrinus* cannot be proved the presence of *Actinocamax* in the absence of *Gonioteuthis* is a useful field guide to the *socialis* Zone in Northern Ireland.

CLOGHASTUCAN CHALK (Variable thickness: see Galboly Chalk)

The base of this unit is marked by a thin zone of wavy bedding with green pebbles and some poorly formed white-skinned flints. The upper limit is taken at the base of a 10-cm belt of wavy-bedded sediment within which there is a sudden increase in the comminuted *Inoceramus* shell component. Cloghastucan Chalk is a flinty unit, in contrast to the underlying *socialis* Zone chalk, the flints again being predominantly silicified burrow-fills of small size. At a variable distance above the base (normally of the order of 2 m) a semicontinuous anastomosing flint up to 17 cm thick— the Oweynamuck Flint Band (Fletcher, *in press*)—occurs within a thin bed defined by well-marked bedding planes. Studies by Fletcher (*in press*) in other areas have shown that the Oweynamuck Flint can be used as a distinctive marker horizon throughout the northern depositional area, and that it can be traced down the east coast as far south as Castle Dobbs in the Carrickfergus (29) Sheet. The basal Cloghastucan Chalk shows evidence of minor penecontemporaneous slumping of incompletely consolidated sediment, for example in the middle slipped section at Oweynamuck. This phenomenon can also be observed in the eastern section at Oweynamuck, near the base of the topmost metre below the Oweynamuck Flint—i.e. the main level of flint development (see Figure 19). Larger-scale slumped bedding is present in the interval between the Oweynamuck Flint and the top of the unit at Dunnaglea, but has not been observed at this level elsewhere.

The fauna is considerably more diverse than that of the *socialis* Zone, but this may be partly a reflection of more favourable collecting conditions. All the species recorded from the underlying zone are present, with the exception of *Uintacrinus*, which exhibits no overlap with *Marsupites*. The zonal index ranges from the base to a level 25 cm below the Oweynamuck Flint, and is represented by both calyx plates and brachials. As in southern England, there is an upward progression from small smooth calyx plates to large plates with strongly developed ornament, the latter occurring in the flinty beds and coinciding with the entry in strength of *Gonioteuthis*. With the exception of the *socialis* Zone elements the recorded fauna of the Cloghastucan Chalk up to the level of the Oweynamuck Flint is as follows:
Neoflabellina cf. '*deltoidea*' (Wedekind)
Retispinopora sp.
Proliserpula ampullacea (J. Sowerby)
Clausa globulosa (d'Orbigny)
Onychocella dichotoma Marsson
indeterminate octocoral
Kingena lima (Defrance)
Orbirhynchia sp.
asteroid marginal ossicles: incl. *Metopaster sp.*,
  Chomataster sp.
Bourgueticrinus papilliformis Griffith & Brydone
B. sp. ['waisted' and elongate columnals]
Marsupites testudinarius (Schlotheim)
Echinocorys sp. small forms near *E. elevata* Griffith & Brydone
Stereocidaris cf. *sceptrifera* (Mantell) [radioles and plates]
Chlamys cretosa (Defrance) forma *nitida* Mantell
Neithea aff. *sexcostata* (S. Woodward)
Plagiostoma? sp.
Pseudoperna boucheroni (sensu Woods *non* Coquand)
Plicatula cf. *barroisi* Peron
Spondylus sp. adnate forms
Gonioteuthis granulata (Blainville)

The distinction between the flat-topped *Echinocorys striata* variants in the underlying member and the pyramidate *elevata* variants in the Cloghastucan Chalk appears to hold good as a criterion for separating the two members in the field (Fletcher, *in press*).

The Cloghastucan fauna compares well with that of the Thanet *Marsupites* Zone despite the marked difference in lithofacies, although it should be noted that certain common Thanet elements, for example the small echinoids *Hagenowia* and *Zeuglopleura*, have not yet been recorded.

The uppermost part of the Cloghastucan Chalk, that is the interval from the Oweynamuck Flint to the upper wavy-bedded horizon, is relatively barren, having yielded only *Inoceramus* chips and *Gonioteuthis*. In the Thanet succession and elsewhere in southern England beds with strongly ornamented *Marsupites* calyx plates and relatively common *Gonioteuthis granulata* are followed by a small thickness of chalk containing plates and brachials of *Uintacrinus anglicus* Rasmussen (see Brydone, 1915; Rasmussen, 1961) which constitutes a kind of 'buffer zone' between the *Marsupites* Zone and the overlying *Offaster pilula* Zone. This crinoid has not yet been found in Ireland, but it is tentatively suggested that this terminal part of the Cloghastucan Chalk corresponds with the beds with *U. anglicus* in southern England.

CREGGAN CHALK   (Thickness—see Galboly Chalk)

This unit is readily recognisable at outcrop by reason of its distinctive lithology, which is more a shell-detrital limestone than a chalk. It is a true '*Inoceramus* Chalk', but with a much higher shell component than in the sediments of the crinoid zones. The base is marked by a sudden increase in comminuted *Inoceramus* shell, coinciding with a level some 10 cm thick of wavy-bedded sediment. *Inoceramus* detritus is present throughout the member, but is most noticeable in the lower beds. In the higher part of the succession minor erosion surfaces marked by concentrations of green pebbles point to conditions of interrupted sedimentation prior to the prolonged period of erosion and mineralisation represented by the Bendoo Pebble Bed (Plates 27 and 30). Small burrow-fill-type flints occur, notably in a belt immediately above and below a well-marked bedding plane in the lower part of the unit.

The hard nature of the sediment inhibits weathering, and fossils are consequently difficult to collect and develop. The lowest beds, immediately above the basal zone of wavy bedding, are relatively slabby due to the presence of numerous thin-shelled oysters aligned parallel to the bedding. Apart from the degree of induration resulting from diagenetic compaction the lithology is closely comparable with the so-called 'Grobfazies' found at this level (*G. granulata–quadrata* Zone) at Lägerdorf in north-west Germany (see Ernst, 1963, stratigraphical section opposite p. 80 and pp. 109–111). The winnowing of the coccolith component in these shell-detrital chalks could perhaps be interpreted as indicative of deposition above the lower limit of wave action (cf. Reid, 1973, p. 361), and it is interesting to note that such chalks can be demonstrated in both Ireland and Germany at this level, and, to a lesser extent, in southern England.

The fauna of the basal slabby beds includes many species ranging up from the underlying members, notably the ubiquitous large lituolid(?) foraminifera, *Neoflabellina sp.*, *Porosphaera globularis* and *Glomerula gordialis*. The oysters are predominantly *Pseudoperna boucheroni*. In addition the following have been recorded:

indeterminate hexactinellid sponges [Port More Borehole]
Orbirhynchia sp.
Stereocidaris sp. radioles [common]
Bourgueticrinus [columnals only: narrow-waisted forms]
Gryphaeostrea cf. *canaliculata* (J Sowerby)
Neithea aff. *sexcostata* (S. Woodward)
pectinid mould: *?Chlamys cretosa* Defrance
Gonioteuthis granulata (Blainville)
lamnid tooth

The higher beds which are relatively poorly fossiliferous have yielded *Echinocorys tectiformis* Griffith & Brydone, together with infrequent small-sized examples of the zonal index. It is probable, however, that *Offaster pilula* is not uncommon at this level, since fragments of thin-tested echinoids are frequently encountered. Belemnites are uncommon and are restricted to *Gonioteuthis* indistinguishable from those of the underlying *Marsupites* chalk. A small *Actinocamax*, from the group of *A. verus* or *A. laevigatus*, was also found at Spy Window [D 330 150] south of Glenarm in east Antrim (Sheet 20), but not in the area of the present sheet. This occurrence in the Northern Ireland Lower Campanian matches the record (Peake and Hancock, 1961, p. 313) of *A. verus* from the 'Zone of *Gonioteuthis*' in Norfolk, although elsewhere in the English Chalk small *Actinocamax* are virtually restricted to the Santonian. In Russia, where these belemnites occur in abundance, the top of the stratigraphical range is the so-called '*Pteria* Beds' of Lower Campanian age (see Christensen *in* Bergstrom and others, 1973, pp. 134–135). A further link with the Russian '*Pteria* Beds' is provided by a specimen of the characteristic fossil—*Oxytoma (Hypoxytoma) tenuicostatum* (Roemer)—in Creggan-type matrix from an unidentified locality in the old Geological Survey of Ireland Collection. This species occurs in the *pilula* Zone of the 'Northern Province' in England, notably at Wells in Norfolk (Peake and Hancock, 1961, p. 313), but is not known from the Chalk of southern England. In Sheet 7 the only horizoned specimen is from the overlying Boheeshane A in the Port More Borehole. The Creggan Chalk of this borehole also yielded poorly preserved small brachiopods (*Orbirhynchia*, *Kingena*), but these were not noted at outcrop.

The pebbles of the Bendoo Pebble Bed are predominantly green-coated (i.e. glauconitised) derived fossils. Extensive collections have been made from this level in the small bay immediately to the east of the Bendoo Plug [D 042 452], where the pebble bed can be examined in plan in the rock-platform. The pebbles are mainly rolled and virtually unrecognisable hexactinellid sponges, but include corroded echinoids and belemnites (*Gonioteuthis*). The better-preserved echinoids are determinable as small examples of *Echinocorys tectiformis* exhibiting features transitional to *E. truncata*; the true diminutive *E. truncata*, characterised by its relatively exaggerated apical area, has not been found. The belemnites, many of which show evidence of attack by clionid sponges and other boring organisms, are for the most part so badly corroded as to be almost indeterminate. Riedel Quotients —a measure of the depth of the pseudalveolus expressed as a ratio to the preserved length of the guard (see Ernst, 1964)—vary between 1/10 and 1/4, suggesting that the belemnite assemblage may span a considerable stratigraphical range, and that the Creggan–Boheeshane non-sequence is significant. It is possible, however, that the forms with a deeper pseudalveolus (*Gonioteuthis quadrata*) may have been collected from sediment of Boheeshane A age, but immediately overlying the pebble bed. Apart from the remanié elements, the matrix of the pebble bed has yielded several small brachiopods, including a possible terebratellid.

The most significant feature of the Sheet 7 (and Northern Ireland) *pilula* Chalk is the fact that not only is it a condensed sequence developed in a bioclastic rather than a chalk lithofacies, but that the greater part of the zonal sequence—interpreted on the basis of the standard southern England successions—is not represented, and was probably never deposited. The echinoid evidence, small-sized *Offaster pilula* and *Echinocorys tectiformis*, permits correlation with the *Echinocorys depressula* Subzone of the southern English *pilula* Zone, but perhaps only with its lower part. The occurrence of *Echinocorys* transitional to the *truncata* form, on the other hand, suggests that the topmost beds may be of 'truncata Belt' age or a little earlier, and equate with a position near the top of the *depressula* Subzone. The non-deposition of higher *pilula* Zone sediment—i.e. the Subzone of abundant *Offaster pilula*—is suggested by the complete absence of *Echinocorys cincta* Brydone, and large forms of the zonal index (including the subspecies *planatus* Brydone [see Ernst, 1971, pp. 194–195, Abb. 18, figs. 7–8] indicative of the uppermost beds of the subzone). The distribution of *Offaster* in the Northern Ireland Chalk, however, is anomalous, since a number of small *O.* cf. *pilula* have now been collected by R. E. H. Reid from Boheeshane A1 in Sheet 7 (personal communication, see below), and a form near to or conspecific with the large subspecies *O. pilula planatus* occurs in abundance at the top of the Upper Campanian Ballintoy Chalk (see p. 97), giving rise to the hitherto unexplained mixed *Offaster pilula*–*Belemnitella mucronata* zonal assemblage recorded by Casey (in Wilson and Robbie, 1966) from Rathlin Island. The low Riedel Quotients of a proportion of the *Gonioteuthis* from basal Boheeshane A1 point to a restricted *quadrata* Zone rather than *pilula* Zone age. *Echinocorys* collected a few metres higher in the succession (Boheeshane A2 *Echinocorys* Bed), however, have been examined by Mr N. B. Peake (Norwich), who has suggested an *Applinocrinus* [*Saccocoma*[*cretaceus*[1] Subzone position on the basis of southern England assemblages. The available evidence, therefore, indicates a significant non-sequence at the Creggan–Boheeshane boundary, with sediments of the *cretaceus* Subzone, or possibly the underlying *Hagenowia* Horizon, resting on an eroded surface approximating to the top of the *depressula* Subzone of the *Offaster pilula* Zone.[2]

BOHEESHANE CHALK (24.25 m thick)

This is a major tripartite member delimited by the Bendoo Pebble Bed below and the erosion surface underlying the base of the Larry Bane Chalk above. It is 24.25 m thick at the type section in Boheeshane Bay (Plate 35). The three units are coded A to C in ascending order.

Boheeshane A (8.36 m) is subdivided into a thin (2.45 m) lower flintless sequence (A1), and a thicker (5.91 m) flinty sequence (A2).

Boheeshane A1, with the exception of occasional discontinuous thin tabular flints intercalated along one bedding plane, is completely flintless, but includes minor surfaces marked by concentrations of green pebbles which point to relatively unstable sedimentary conditions continuing after the episode represented by the Creggan–Boheeshane non-sequence. It can be best examined in the small bay immediately east of the Bendoo Plug, but it is difficult to gain access to this locality. The occurrence of small *Offaster* below the erosion surfaces marked 'a' and 'b' on the fallen block sections east of Port Braddan and the White Park Bay Waterfall (see Figure 19, Sections F and D) is noteworthy, since the *Gonioteuthis* and *Echinocorys* point to a restricted *quadrata* Zone age for this post-Creggan flintless unit. A comparable situation is found in southern England, where Brydone (1912) referred to small *Offaster* (in addition to a thin-tested 'dwarf form') from the *quadrata* Zone of Hants, although no stratigraphical details were given.

The remaining macrofauna includes:
sponges, including *Rhizopoterion sp.*
Conorca turbinella (J. Sowerby)
Kingena sp. [small: Port More Borehole]

Orbirhynchia sp. [Port More Borehole]
Echinocorys [small pyramidate forms predominate, together with rounded inflated forms similar to those of A2]
pectinid indet. [Port More Borehole]
Gonioteuthis [relatively common; predominantly slim forms with low Riedel Quotients]
fish remains, not identified

Boheeshane A2 is essentially thin-bedded, with tabular flints developing along the more significant bedding planes. These tabular flints may be absent locally, while elsewhere they may be strongly represented. Splitting of tabular courses over a few metres can be observed in the type section, particularly at the top, where there is an easily recognisable belt of close-set tabulars. A2 contains in addition several levels with small nodular burrow-fill flints, and a large flattened flint is developed about 2 m above the base, immediately above the *Belemnitella praecursor* Bed.

There are two important fossil horizons:

1 About 1 m above the base *Echinocorys* is relatively common, the assemblage comprising predominantly medium-sized rounded forms with a wide ellipsoidal base, together with subordinate small pyramidate forms of Boheeshane A1 type. This assemblage has been tentatively attributed by N. B. Peake to the *Hagenowia* Horizon.

2 The *Belemnitella praecursor* Bed, which is most conveniently examined in the shore-platform at Port Braddan (Plate 28), and in a bluff in the small bay east of the Bendoo Plug, is a thin bed characterised by abundant large *Belemnitella* ex. gr. *praecursor* Stolley, which dominate the belemnite assemblage to the virtual exclusion of the *Gonioteuthis* lineage. *Belemnitella* has also been noted from other horizons within A2, and it may therefore be more widely distributed than realised at present. English *Belemnitella* of this type were figured by Jeletzky (1955, pl. 57, figs. 3a–b, from the former Southampton Waterworks pit south of Shawford, Hants—Brydone's Hants locality no. 1086: see Brydone, 1912, pp. 10, 100—and pl. 58, figs. 1a–c, from Bramford near Ipswich— erroneously stated to be from the junction of the *quadrata* and *mucronata* zones). Markham (1971) reported, in fact, that a unit characterised by *Belemnitella praecursor* is present in the *quadrata* Zone of the Ipswich area, and there is evidence of the same horizon at Stiffkey in north Norfolk (Brydone, 1933, p. 290; current IGS fieldwork). Further afield, the *Belemnitella*/*Gonioteuthis* mixed assemblage of the 'Smectite' at Hallembaye, Belgium, is probably also of this age (see Schmid, 1967, p. 464). Throughout the Santonian and Lower Campanian of the North European Province *Belemnitella* (together with *Actinocamax* and larger actinocamaxoid belemnites) occupied a distinct belemnite subprovince [the Central Russian Subprovince of Christensen, 1975] situated to the north and east of a subprovince characterised by *Gonioteuthis* [the Central European Subprovince (Christensen, 1975)], with the essentially northern *Belemnitella* progressively extending its area of distribution southwards (see Naidin, 1960; Christensen, 1975, pp. 26–29; Stevens, 1973, p. 394). In this larger context the displacement of *Gonioteuthis* by *Belemnitella* in the lower part of the *quadrata* Zone in Britain and adjacent areas appears to be an event of considerable biostratigraphical significance, possibly reflecting the temporary incursion of the Central Russian fauna into the *Gonioteuthis* subprovince.

Apart from the species mentioned above, the following macrofauna has been recorded from Boheeshane A2 as a whole:
Ditrupa (Ditrupa) sp.
Flucticularia sp.
Coelosmilia spp. including *C. regularis* Tomes
Kingena cf. *blackmorei* Owen [Port More Borehole only]
Terebratulina ex gr. *striatula*–*chrysalis* [Port More Borehole only]
Micraster sp. juv. cf. *schroederi* Stolley
Stereocidaris 'serrifera' Auctt.: radioles
Gryphaeostrea sp.

---

1 Peck (1973) demonstrated that *Saccocoma* strictly applies only to the Kimmeridgian forms, and introduced *Applinocrinus* for the Upper Cretaceous forms, thereby necessitating a change in the subzonal nomenclature.
2 For a definition of the current subdivision of the English *quadrata* Zone see Peake (in Hancock, 1972, pp. 60, 79–80, and 113–115).

Oxytoma (*Hypoxytoma*) *tenuicostatum* (Roemer)
Gonioteuthis [diverse forms grouped around *G. quadrata* but including some apparently primitive forms with relatively shallow pseudalveolus]

Boheeshane A2 is characterised by *Coelosmilia regularis* and *Belemnitella* ex gr. *praecursor*, and possible also by *Kingena blackmorei*, although only one questionable specimen of the latter species has been found so far. *C. regularis* was originally described from the Lower Campanian (probably *quadrata* Zone) of the Salisbury area (Tomes, 1899), but it also occurs at various localities in north-west Germany, for example, at Höver, south of Hannover [C. J. W.; unpublished observation]. In Sheet 7 it has been collected at outcrop in Boheeshane Bay, and from the Port More Borehole. The occurrence of this coral in Boheeshane A2 and correlative horizons in other areas may prove to be an important stratigraphical marker. The large kingenid *K. blackmorei* (see Owen, 1970, pl. 7, figs. 1–3; pp. 63–64) is an element of the fauna at Brydone's locality 1086, and it is interesting to note the probable extension of its geographical range to Northern Ireland.

The *Belemnitella* occurrence in Boheeshane A2 must not be confused with the occurrence of large *Belemnitella* ex gr. *praecursor* in Boheeshane C1 and particularly C2 (see p. 95)—i.e. immediately above and below the boundary between the Lower and Upper Campanian as defined by the (local) extinction of *Gonioteuthis*. The presence of *Belemnitella* is not in itself indicative of the *Belemnitella mucronata* Zone, and considerable confusion has arisen in the past from failure to appreciate this point (e.g. Brydone, 1912, p. 10). It is clear that there are two horizons with *Belemnitella* in the *quadrata* Zone: the lower, Boheeshane A2/Smectite horizon; and the higher, Boheeshane C horizon, which corresponds to the 'Overlap Zone' of Continental workers (see Schmid, 1967), and which is probably also represented in southern England. *Belemnitella* is also known as an extreme rarity from the *pilula* Zone in Hants, but the relevant part of the succession is missing in Northern Ireland at the Creggan–Boheeshane non-sequence.

Boheeshane B (8.69 m) is subdivided at a large and particularly conspicuous flint 3.9 m above the base into two parts, coded B1 and B2 in ascending order. The upper limit is marked by a bedding plane immediately below the flinty zone at the base of Boheeshane C. Boheeshane B appears to exhibit some lateral variation, with reductions in thickness to east and west of the type section, and is particularly difficult to recognise in the Port More Borehole. Although an extensive mesofauna has been collected from air-weathered sections, notably the Dunnaglea cliffs and Portbraddan, the larger and more diagnostic fossils are rather poorly represented. It must be emphasised however, that there are no localities where this part of the succession is exposed as bedding surfaces, and the faunal lists may therefore be somewhat unrepresentative.

Boheeshane B1 is readily recognised in the field by an *Orbirhynchia* Band about 1.5 m thick [with *O. bella* Pettitt] which occurs in the middle part of the succession, and which terminates at a well-marked bedding plane along which a tabular flint is developed locally, for example near the base of the cliff at Port Braddan (Plates 28 and 29). The brachiopods occur throughout but are concentrated in the top 15 cm. The *Orbirhynchia* Band is a particularly important marker in successions where it has not proved possible to differentiate the various subdivisions of the Boheeshane Chalk, as in the case of the apparently expanded sequence of Church Bay (Rathlin Island) in the Ballycastle (8) Sheet, and also in the East Antrim depositional area, notably at Creggan near Glenarm. It was present in the Port More Borehole, although the exact depth is somewhat uncertain (see p. 113). It is interesting to note that *Orbirhynchia* is well represented in the brachiopod fauna from Brydone's Hants locality 1086 (IGS collection), but it is not known whether it occurred at one horizon only, or whether it was distributed throughout the succession.

The following fauna has been recorded from B1:

Porosphaera globularis
P. patelliformis Hinde
Retispinopora arbusculum Brydone
Coelosmilia sp. juv.
'*Parasmilia*' *cylindrica* Milne Edwards & Haime
Conorca turbinella
Ditrupa sp.
Glomerula gordialis
Proliserpula ampullacea
Lunulites?
Meliceritites dollfussi
Onychocella dichotoma (Marsson)
cf. *Cretirhynchia intermedia* Pettitt
Kingena sp. juv.: *blackmorei*? Owen
Orbirhynchia bella Pettitt
Isocrinus? columnal
Applinocrinus calyx? [a somewhat uncertain determination, recorded by Wood *in* Wilson (1972) as *Saccocoma* calyx: possibly an ophiuroid ossicle]
Ophiomusium granulosum (Roemer) [isolated ossicles are relatively common on weathered surfaces]
Ophiura cf. *substriata* Rasmussen
Echinocorys [small high variant (e.g. CJW 1239) with narrow base agreeing with specimens in the IGS collection (Zf 2324–5) from the *quadrata* Zone of South Charford, Hants, some 10 km south of Salisbury. A large, inflated form is also present, but is so far known only from an incomplete specimen]
Micraster cf. *schroederi* Stolley
Stereocidaris '*serrifera*' Auctt.: radioles
asteroid marginal ossicles, including *Pycinaster* sp.
Gonioteuthis of the *quadrata* evolutionary stage

Boheeshane B2 is the lowest unit shown on the vertical range chart (Figure 20) for the upper part of the pre-North Antrim Hardground succession. Practically all the data have been taken from Larry Bane Bay, where the unit is readily accessible for collecting. The sudden increase in diversity of the brachiopod fauna is particularly striking, this being a faunal characteristic which persists to the top of the Larry Bane Chalk. The Boheeshane B2 fauna can be regarded, in fact, as a transition fauna to the true *mucronata* Zone fauna which enters in the middle of the overlying Boheeshane C.

There are three important fossil horizons:

1 The 1st *Galeola* Bed. This is the lowest of three well-marked *Galeola* horizons in the Irish succession, although occasional *Galeola* are also found at other levels. It is well seen in Larry Bane Bay, at a level 2.15 m below the top of the unit, and it is also present on the rock platform in Church Bay, Rathlin Island. Due to the intensely hard matrix at both localities it has not been possible to develop specimens satisfactorily, but, to judge from the stratigraphical position, it is probable that the species in question is *G. papillosa* Klein sensu Ernst (Ernst, 1971, pl. 22, figs. 2–4).

2 The depressed *Echinocorys* Bed. This horizon has been observed in Larry Bane Bay, where it underlies the 1st *Galeola* Bed, and also in the White Park Bay amphitheatre section (perhaps at a slightly higher level) and Church Bay (Rathlin Island) beach platform. It is characterised by a large and very depressed *Echinocorys* [e.g. CJW 1245, 2366], which is apparently undescribed and which is not known from the *quadrata* Zone of England (Peake, personal communication). Poorly preserved (purple) sponge meshworks are conspicuous immediately below the bed in the western part of Larry Bane Bay.

3 The topmost 0.5 m of Boheeshane B2 yields a precursor of the *Echinocorys conica* group ['*preconica*', Peake MS] such as characterises

the relevant part of the succession at Roger's Whitening Pit (Brydone, Hants loc. 1153), and the lowest true examples of the *conica* group.

With the exception of the species shown in the range chart, the following fauna has been recorded:

Porosphaera globularis
Conorca turbinella
Onychocella cf. *gibbosa* (Marsson)
Pavolunulites scandens Brydone
Terebratulina cf. *chrysalis* (Schlotheim)
Echinocorys [large inflated forms of uncertain affinity]
cidarid radiole
Gonioteuthis [very few determinable specimens have been collected from the hard matrix. The assemblage includes small forms which are probably referable to the subspecies *G. quadrata gracilis*—i.e. the terminal retrogressive evolutionary stage of the *Gonioteuthis westfalica–granulata–quadrata–quadrata gracilis* lineage (see Ernst, 1964, pl. 2, figs. 7–10, and discussion in text)]

Boheeshane C (7.21 m), the top unit of the Boheeshane Chalk is delimited by a basal bedding plane (locally relatively well marked, as, for example, in the type section, but elsewhere indistinct) and by the base of the overlying Larry Bane Chalk. Within the unit a number of distinctive flint courses can be recognised (Fletcher, *in press*), comprising a basal zone of flints, two massive flints occupying a subcentral and central position respectively, and two further flinty zones in the upper part of the succession. The higher of the two massive flints (the Whitehead Flint Band) corresponds with the (local) upper limit of the range of *Gonioteuthis*, and consequently with the Lower–Upper Campanian boundary at the junction of the *quadrata* and *mucronata* Zones. The other flints have been named, in ascending order, Basal Zone, Basal Massive, Lower Top Flinty Zone and Upper Top Flinty Zone, and they can be recognised wherever Boheeshane C is exposed. Boheeshane C is, in fact, the most constant of all the Boheeshane subdivisions, retaining its lithostratigraphical character even into areas (e.g. the eastern depositional basin) where Boheeshane A and B cannot be clearly differentiated. The Whitehead Flint serves to divide Boheeshane C into a lower *quadrata* portion (C1) and a higher *mucronata* Zone portion (C2) respectively. Boheeshane C is richly fossiliferous throughout, and an extensive fauna has been collected from Larry Bane Bay, where every part of the succession is easily accessible on the rock-platform, and Portbraddan, where the top part of C2 is seen at beach level at the north-western end of the section.

There are no particularly significant fossil horizons in Boheeshane C1, although the bed immediately below the Basal Massive Flint yields a rich brachiopod fauna. It is probable that the fauna is more diverse than recorded here, since, with the exception of a bluff by the path leading to the bay east of the Bendoo Plug, there are no air-weathered surfaces available for collecting. At this latter locality a superb example of *Isocrania paucicostata* (Bosquet) was collected at the level of the Basal Flinty Zone, together with disciform bryozoa belonging to the genus *Lunulites*; the latter occur at the same horizon in other areas, e.g. in east Antrim, and provide a good guide to the base of Boheeshane C.

The recorded fauna, excluding species on the range chart, is as follows:

Porosphaera globularis
Retispinopora cf. *arbusculum* [Bendoo]
Lunulites martini Voigt
L. semilunaris (Hagenow)
Pavolunulites scandens
Conorca turbinella
Isocrania paucicostata
asteroid marginal: *Recurvaster?*

Echinocorys ex gr. *conica*
E. 'pre-conica' Peake MS
E. cf. *subglobosa*
Spondylus cf. *spinosus* (J. Sowerby) [1″ Sheet 20 only]
Belemnitella sp. indet. [uncommon]
Gonioteuthis [small forms, presumed to be the terminal stage of the *Gonioteuthis* lineage, that is *G. quadrata gracilis*]

The joint occurrence of *Gonioteuthis* and *Belemnitella* in Boheeshane C1 constitutes an 'overlap-zone' such as occurs on the Continent in the highest part of the Lower Campanian. It is uncertain, however, whether this is synchronous with the Boheeshane overlap zone, since there are serious discrepancies in the vertical ranges and frequency maxima of key species in the accompanying fauna (see discussion on p. 94). The rarity of belemnites in Boheeshane C1 contrasts with their relative abundance in C2.

In Boheeshane C2 three important fossil horizons can be recognised:

1 The interval from the Whitehead Flint to the base of the Lower Top Flinty Zone is characterised by large elongate *Belemnitella* tentatively referred to the *praecursor* group. These belemnites exhibit small fissure-angles, and the protoconch–apex distance is large relative to the dorsoventral diameter at the protoconch; in this latter respect the Boheeshane C2 belemnites are readily distinguishable from the belemnites of the overlying Larry Bane Chalk. The degree of vascularisation of the ventral surface of the guard is reminiscent of belemnites of the group of *B. mucronata senior* Nowak, and it is possible that these high Boheeshane belemnites are forms morphologically transitional between the essentially Lower Campanian *praecursor* group and *Belemnitella mucronata* sensu lato of the Upper Campanian. The Boheeshane C2 assemblage compares well with material from Roger's Whitening Pit (Brydone's Hants locality no. 1153) (IGS collection). Medium-sized *Micraster* ex gr. *schroederi–glyphus* were collected from this interval in Larry Bane Bay, but not noted elsewhere.

2 The large (or 2nd) *Galeola* Bed, which is more or less coextensive with the Upper Top Flinty Zone. This bed marks the lowest recorded occurrence of the large rhynchonellid *Cretirhynchia woodwardi* (Davidson) but is particularly characterised by a size-maximum in *Galeola*. The *Galeola* assemblage includes forms figured by Ernst as *G. papillosa papillosa* and *G. papillosa basiplana* (1971, pl. 23, figs. 1a–c, 8–9). Comparable forms occur in southern England (Roger's Whitening Pit and Farlington Redoubt, Hants) at an equivalent level, and are also known on the Continent from the Craie d'Obourg of Harmignies, Belgium; from the Lower–Upper Campanian boundary succession of the Misburg area, West Germany; and also from the Vistula Valley north of Kraków, Poland [C. J. W.; unpublished observation based on material collected by Dr R. B. Stokes]. The large *Galeola* Bed has been traced from eastern Antrim (Larne area) to the north coast, and is particularly accessible in Larry Bane Bay, and in Church Bay (Rathlin Island). It is clearly a most significant marker horizon in Northern Ireland, and to judge from occurrences of the characteristic echinoid assemblage elsewhere, it may prove to be a key to the interpretation of Lower–Upper Campanian boundary successions throughout north-western Europe (see p. 110).

3 The uppermost 5 cm of Boheeshane C, in common with other beds higher in the succession immediately below the junction between two major sedimentary units, is richly fossiliferous (see range chart, Figure 23). This bed appears to mark a temporary phase of shallowing prior to the period of erosion which preceded the deposition of the Larry Bane Chalk, during which a diverse echinoid biotope developed; locally, incipient hardground features are present, for example intense induration accompanied by pink colouration resulting from mineralisation; the preservation of hexactinellid sponges as limonitic pseudomorphs. The echinoid

assemblage comprises predominantly *Echinocorys* ex gr. *conica* and the giant *E. subglobosa* (Lambert, 1903, pl. 2, figs. 7, 8) with *Cardiotaxis* ex gr. *heberti*, small *Micraster* ex gr. *schroederi–glyphus* and *M.* (*Isomicraster*) cf. *stolleyi* as subordinate elements. *Micraster gibbus* sensu Ernst 1970 = *M. fastigatus* Gauthier, which is so abundant at this level in Germany and elsewhere, has not been found in Northern Ireland. It should be noted that the terminal Boheeshane C is the only horizon in the Irish *mucronata* Zone which has yielded *Micraster* (*Isomicraster*); this extremely limited distribution contrasts with the somewhat more frequent occurrence of *M.* (*I.*) *stolleyi* at several levels in the lower half of the Norfolk succession—that is in the equivalent of the Ballintoy and lower Glenarm Chalk.

This bed is particularly accessible for collecting at the northern end of the Portbraddan section (Plate 31), where the overlying Larry Bane Chalk descends to the shore-platform. Outside Sheet 7 it can be examined near Retreat Castle in the Cushendall (14) Sheet, where the top of Boheeshane C forms the floor of an old quarry and contains numerous large pachydiscid ammonites.

An extensive fauna has been collected from Boheeshane C2, both in Sheet 7 and in east Antrim. Supplementary records from outside the sheet are marked by an asterisk.

Porosphaera globularis
**Pavolunulites scandens*
Ditrupa sp. [circular transverse section]
Flucticularia sp.
*asteroid marginals including *Metopaster* cf. *tumidus* and *Ophryaster*?
Galerites sp.
Inoceramus, not determined
*Limea granulata
Chlamys mantelliana (d'Orbigny)
**Neithea sexcostata* (S. Woodward)
Pycnodonte vesicularis (Lamarck)
Spondylus dutempleanus d'Orbigny
**S.* cf. *spinosus*
Belemnitella spp.: large elongate forms ?ex gr. *praecursor* (see p. 95), and subordinate less elongate forms near *B. mucronata senior* Nowak.
small pachydiscid ammonite indet. [Port Braddan]
fish remains

Despite prolonged search, *Gonioteuthis* has not been observed above the Whitehead Flint—i.e. above the level taken as the Lower–Upper Campanian boundary. In Germany, however, the *Gonioteuthis/Belemnitella* 'overlap zone' extends up as high as the beds with frequency-maxima of *Echinocorys* ex gr. *conica* (Khosrovschahian, unpublished thesis, 1972) equivalent to the Larry Bane Chalk of the Irish succession. It is therefore possible that the extinction of *Gonioteuthis* was not synchronous across the European area—as Jeletzky (1951, p. 127) considered—and that the boundary between the Lower and Upper Campanian has been drawn at different levels in different areas. It must be understood that the Lower Campanian–Upper Campanian boundary as drawn in Northern Ireland has no international correlative status, and is determined solely on the (local) extinction of *Gonioteuthis*. It remains to be seen whether the level of size-maximum in *Galeola* and/or the levels with frequency-maxima of *Echinocorys* ex gr. *conica* will prove a more reliable biostratigraphical marker for the boundary of the substages. Certainly on present echinoid evidence it is probable that the extinction of *Gonioteuthis* is diachronous, occurring progressively later from west to east.

LARRY BANE CHALK (7.3 m)

The Larry Bane Chalk comprises two thin units (coded A and B respectively) having a thickness ratio of almost exactly 1:2 and delimited by erosional separation planes forming cliff notches in weathered sections. These limiting surfaces are marked by strong negative peaks in the resistivity profile of the Port More Borehole. The Larry Bane unit would thus provide a key marker in the interpretation of other borehole successions. The type section is taken in the cliff at Larry Bane Head [D 049 451], where the unit is 7.3 m thick (Plates 33 and 34).

The Larry Bane fauna is characterised by *Cretirhynchia woodwardi* and *Kingena pentangulata*, together with large corpulent and strongly vascularised *Belemnitella* of the group of *B. mucronata senior* Nowak (Nowak, 1913, pl. 42, fig. 22). The Larry Bane belemnite assemblage is clearly distinct from that of the preceding Boheeshane C2: the belemnites are more strongly vascularised, and exhibit larger fissure-angles and lesser elongation from the protoconch to the apex. *Echinocorys*, including large forms of the *marginata–subglobosa* group and forms grouped around *E. conica*, occur throughout, with the latter concentrated in the uppermost 10 cm of A and B respectively. The terminal Larry Bane B echinoid-bed assemblage comprises almost exclusively *Echinocorys* ex gr. *conica*, with infrequent *Galeola*, and represents the frequency-maximum for the *conica* group in the Irish succession, and the possible correlative of the top of the *Gonioteuthis/Belemnitella* 'overlap zone' of Continental successions. Unlike the majority of echinoids in the Irish Chalk, the echinoids from this bed can be developed clean from their matrix, this perhaps being a function of early-stage lithification inhibiting pressure-solution during the phase of sediment compaction (see Wolfe, 1968). The assemblage is morphologically heterogeneous and includes many of the subspecies established by Lambert (1903) and Smiser (1935) from the Belgian Chalk. A similar situation has recently been reported from the top of the Lower Campanian of the Misburg area, West Germany (Khosrovschahian, 1972), and it appears likely that a statistical analysis of the *conica* group would demonstrate high population variability, rather than actual geographical subspeciation.

Faunas supplementary to those recorded in the range chart are as follows, with records from outside Sheet 7 indicated by an asterisk:

FROM LARRY BANE A:
Porosphaera globularis
**Pavolunulites richteri* (Hagenow)
*'*Vincularia*' *sp.*
**Conorca turbinella*
Magas plena (Katz) [common in the Weybourne Chalk of Norfolk]
Terebratulina cf. *chrysalis*
*asteroid marginals including *Metopaster* or *Recurvaster*, *Teichaster*?

FROM LARRY BANE B:
Porosphaera globularis
P. sp. [patelliform]
**Meliceritites propinqua* (Marsson)
**Pavolunulites richteri*
**Glomerula gordialis*
Sclerostyla macropus (J. Sowerby)
*miscellaneous small serpulids including a 7-angled form, and a smooth spiral form
cf. *Chlamys cretosa* (Defrance)
* ?pectinid fragment: ?'*Aequipecten*' aff. *sarumensis* Woods [undescribed Norwich *mucronata* Zone species]

BALLINTOY CHALK (12.62 m)

The Ballintoy Chalk comprises two well defined units with a thickness ratio of 3:1 (Ballintoy A and B), of which the lower is almost exactly twice the thickness of the underlying Larry Bane B. The member is terminated by the Altachuile Breccia (Fletcher, *in press*), which consists of a belt of wavy-bedded sediment containing pellets of indurated chalk and scattered angular chips of flint enclosed within a chalk matrix. This key stratigraphical marker, which can be traced from east Antrim to the north coast, was interpreted by Fletcher as the result of wave action on partially consolidated sediment during a period of violent weather conditions such as a hurricane. What is of importance is the occurrence of flint clasts *within* a flinty succession, indicating that some at least of the flint was already fully formed *during* chalk deposition, rather than during subsequent diagenesis. The disturbed zone reaches its maximum intensity in the area of Rathlin Island (Altachuile Bay), but it can be examined in Sheet 7 in the cliff nearest the road at Portbraddan, and at Ballintoy Harbour in the vicinity of the blowholes at the seaward end of the eastern part of the rock platform. The Altachuile Breccia yields large *Offaster* in abundance.

Ballintoy A totals 9.62 m, of which the bottom 3.6 m (A1) is characterised by large variants of *Echinocorys* ex gr. *conica*, including the angulate form *E. lamberti* Smiser (Lambert, 1903, pl. 4, fig. 3), and small examples of the undescribed pyramidate forms which occur in the higher part of the Eaton Chalk subdivision of the Norwich *mucronata* Zone (Peake and Hancock, 1961). Ballintoy A1 marks the culmination of the range of *Echinocorys* ex gr. *conica*, which dominates the fauna to the virtual exclusion of other fossils. Belemnites and brachiopods, in particular, are distinctly uncommon at this level, in marked contrast to their relative abundance in the underlying Larry Bane and Boheeshane units. *Pycnodonte (Phygraea) vesicularis* and a questionable *Pseudoptera coerulescens* (Nilsson) were recorded from the Port More Borehole.

The remainder of Ballintoy A (A2) marks the return of a rich belemnite fauna—which is still dominated by large *Belemnitella* of the *senior* group—together with a more diverse benthos. Two fossil horizons can be noted:

1 The 3rd *Galeola* Bed, approximately 1 m below the top of A2. This bed marks the upper limit of the range of *Belemnitella* ex gr. *senior*, and yields abundant small and medium-sized *Galeola*. The *Galeola* assemblage is dominated by *G. papillosa basiplana* Ernst (Ernst, 1971, pl. 23, figs. 2–6), but includes depressed variants which may be referable to the subspecies *G. papillosa corcula* (Goldfuss)—see Ernst (1971, pp. 217–218; pl. 23, figs. 7a–c). The dominant form is Rowe's *Echinocorys bayfieldi* (MS name), which he collected from a bed in the lower part of the *mucronata* Zone exposed in pits around Norwich, together with a small *Galerites*[1] (*Conulus minor* Rowe MS). More recent collecting (Ernst, 1971, pp. 176, 220) has demonstrated that *G. papillosa basiplana* is relatively abundant in Bed Z of the Weybourne coast section (see Peake and Hancock, 1961, fig. 6) at the junction of the Eaton and Weybourne Chalk subdivisions of the Norfolk *mucronata* Zone. At this locality it is associated with a small *Galerites*—referred by Ernst to *G. sulcatoradiatus* (Goldfuss)—which may be Rowe's *Conulus minor*. This form has also been collected from Ballintoy A2, but not from the 3rd *Galeola* Bed.

If it is possible to use echinoid frequency maxima in different areas as correlatable 'biostratigraphic events'—cf. the 'pulse faunas' demonstrated by Jeans (1968) in the Lower Chalk of southern England—it is tempting to equate Ballintoy A with the Eaton Chalk of the Peake and Hancock classification. In both areas beds with a frequency-maximum of *Galeola papillosa basiplana* succeed beds yielding large forms of the *Echinocorys conica* group. Peake and Hancock did not define the lower limit of their Eaton Chalk, so that it is not clear whether this should include part or all of the Larry Bane Chalk of Northern Ireland. It is interesting to note, however, that the main belt of *Echinocorys* ex gr. *conica* in Norfolk, e.g. Tharston (Rowe's Norfolk locality no. 49) and Newfound Farm, occurs stratigraphically *below* beds with *Galeola maxima*.

2 The ammonite beds. The interval between the 3rd *Galeola* Bed and the top of Ballintoy A2 is characterised by large pachydiscid ammonites which, by comparison with occurrences in Germany, are probably referable to *Patagiosites? [Pachydiscus] stobaei* (Nilsson). These ammonites are particularly well known from Ballintoy Harbour, where they are seen on bedding-planes near the area where the nets are dried.

Supplementary faunas from A2 include:
Porosphaera globularis
Coelosmilia laxa?
**Conorca turbinella*
Ditrupa (Pentaditrupa) sp.
Magas sp.
Terebratulina ex gr. *chrysalis*
Echinocorys: including *E. gibba oviformis* Lambert (Lambert, 1903, pl. 2, figs. 3–5) and large unidentified globose variants. A distinctive depressed form occurs commonly in the top 10 cm.
Phymosoma sp.
Plicatula hantonensis Brydone

The small rhynchonellid *Cretirhynchia lentiformis* is particularly characteristic of this unit, occurring relatively frequently in the third *Galeola* Bed—where it is associated with *C. woodwardi*—and in the uppermost 10 cm.

Ballintoy A2, characterised by abundant *Galeola papillosa basiplana* and giant pachydiscids, clearly corresponds to the *Pachydiscus stobaei/G. papillosa basiplana* Subzone of the German standard low Upper Campanian section at Misburg near Hanover, and correlates approximately with the *stratum typicum* of the neotype population of *Belemnitella mucronata* proposed by Christensen and others (1975). It is interesting to note, however, that the *Belemnitella* assemblage at this level in Ireland still comprises predominantly corpulent forms referable to the group of *B. mucronata senior*.

Ballintoy B (3 m) The upper bed of the Ballintoy Chalk is magnificently exposed in clean air-weathered sections in St Thomas' Cliffs, Rathlin Island (Sheet 8), but within the area of Sheet 7 the exposures are less satisfactory or occur in a relatively inaccessible position in the cliff. The bed can be examined best at various points on the rock platform at Ballintoy Harbour, and it forms the base of the cliff in the Sliddery Cove section, east of Portrush. It is almost certainly this level which was referred to in the description of faunas from St Thomas' Cliffs and Cooraghy in the Ballycastle Memoir (Wilson and Robbie, 1966, p. 169). In these sections it can be observed that the fauna is relatively poor, the only common fossils being calcisponges and bryozoa; the larger fossils, such as echinoids, tend to be crushed and/or fragmented. *Offaster pilula* indistinguishable from the smaller members of the *planoconvexus* Bed assemblage—i.e. *O. pilula planatus* Brydone—at the top of the southern English *pilula* Zone enters halfway up the bed, and reaches a frequency-maximum in the Altachuile Breccia. This anomalous occurrence of true *Offaster* in the Upper Campanian is the only well-attested record at this level anywhere in Europe, although Brydone (1912, p. 111) indicated that they were known to him from post-*quadrata* Zone Chalk, albeit as a rarity, and a questionable example from the Chalk of 'Catton' is preserved in the Brydone Collection at Norwich Castle Museum. The *Offaster* line continued, in fact, into the Maastrichtian, since Kutscher has recently found specimens in the upper part of the Lower Maastrichtian of Rügen, G. D. R. (Ernst, 1971, footnote on p. 206). The interesting feature about the Ballintoy *Offaster* fauna is that it may represent a late-stage evolutionary divergence

---

[1] This small *Galerites* is probably the same form figured by Lambert (1911, pl. 2, figs. 4–8) as *Echinoconus hannoniensis* from Harmignies, Belgium.

away from the main (burrowing) *Offaster* line towards the epibenthonic *Galeola* subsequent to the evolution of the *Galeola senonensis–papillosa–papillosa basiplana* lineage.

Supplementary faunas from Ballintoy B are listed below, with records from Rathlin denoted by an asterisk:
*calcisponge, not determined
*Porosphaera globularis
*Pavolunulites scandens
*Conorca turbinella
*Isocrania paucicostata
Terebratulina chrysalis
Echinocorys: including *E. gibba oviformis* and *E. ovata* Lambert sensu Smiser (see Smiser, 1935, Figures 5a–c)
Phymosoma sp.
Galerites sp.: ?*E. sulcatoradiatus* sensu Ernst
*Limatula aff. decussata (Goldfuss)

Ballintoy B marks the top of the range of *Cretirhynchia lentiformis*, and the replacement of the corpulent *Belemnitella* of the *senior* group by slimmer forms, that is *Belemnitella mucronata* s.s. as interpreted by Christensen and others (1975). Brachiopods are distinctly uncommon.

Glenarm Chalk (6.19 m thick)

The type section for this member is taken in the vicinity of Glenarm in east Antrim where four units, coded A to D, have been recognised. In the north Antrim depositional basin, however, the correlative unit includes a closely spaced pair of hardgrounds, the upper of which is much more strongly developed than the lower. These hardgrounds are grouped together as the North Antrim Hardground[1], the prefix North Antrim being used to distinguish this double hardground from the South Antrim Hardgrounds—a complex of four or more erosion surfaces—which is developed higher in the succession, but with which it has been confused in the past (see Wood *in* Wilson, 1972, pp. 57–58, fig. 22). The North Antrim Hardground in fact marks a level of major faunal and sedimentological change which can be correlated to a greater or lesser extent throughout northern Europe, and which corresponds to the junction of the Weybourne and Beeston subdivisions of the Norfolk succession. For this reason the Glenarm Chalk of Sheet 7 is divided into two units, coded α and β, which reflect this faunal change, the line of demarcation being taken at the top of the main hardground. The sedimentological change at the North Antrim Hardground is expressed as a higher proportion of bioclastic debris in the chalk, and is detectable in the resistivity profile of the Port More Borehole as a pronounced negative shift. In addition, flints in the post-North Antrim Hardground succession up to the top of the Ballymagarry Chalk are predominantly rounded massive forms, contrasting with the smaller burrow-fill-type flints of the underlying succession.

Lithologically Glenarm α (4.62 m at Larry Bane Quarry) comprises two flinty belts separated by a flintless interval; the lowest 1 m of the bed contains three well developed wavy bedding-planes.

The basal thin-bedded sediment yields a rich belemnite fauna, and can be examined to advantage in the shore platform at Ballintoy Harbour, and in the cliffs of Sliddery Cove west of Portrush. At the latter locality the sediment is markedly less indurated than in areas to the east, and the belemnites can be extracted without difficulty. *Offaster pilula planatus* continues from Ballintoy B, but has not been found more than 30 cm above the base. The belemnites are predominantly small elongate forms, referable to *Belemnitella mucronata* sensu stricto. The base of the upper flinty belt can be distinguished as an *Echinocorys* Bed, which has been recognised in both Ballintoy Harbour and Sliddery Cove; at the former locality it can be observed in plan in the vicinity of the blowhole in the eastern half of the chalk platform, where numerous *E. cf. ovata* sensu Smiser have been collected. This bed has also yielded the highest examples of *Cardiotaxis ex gr. heberti*. The *Echinocorys* fauna of the higher part of Glenarm α is somewhat more diverse, including, in addition to *E. cf. ovata* [sensu Smiser], late records of *E. gibba oviformis* and small, apparently undescribed, narrow-based forms. CJW 1878 from 70 cm below the North Antrim Hardground has been attributed by Peake to an undescribed globose variant which occurs in a hardground approximately 1 m above the main Catton Sponge Bed in the Norfolk succession (see Peake and Hancock, 1970, p. 339E). The low occurrence of this variant in the Northern Ireland succession is of interest since the Catton Sponge Bed and North Antrim Hardground appear to reflect the same sedimentological event. The upper flinty belt includes a level (Sliddery Cove) of small *Galerites cf. vulgaris* incorporated in one flint course; treatment of these flints with acid produces well-preserved undistorted flint steinkerns. *Micraster ex gr. schroederi–glyphus* was also found here, this being the highest record in the succession. At the top of Glenarm α there are two hardgrounds some 40 cm apart forming the North Antrim Hardground complex. The lower of the two is little more than a wavy bedding plane accentuated by weak green colouration due presumably to glauconitisation. *Echinocorys* occurs in the intervening chalk, but cannot be extracted from the matrix. The terminal hardground is heavily mineralised, and appears in section as an irregular dark green line. The erosion surface is relatively flat in contrast to the mushroom-like residual structures which are developed in other Chalk hardgrounds (e.g. the Chalk Rock), and is overlain by a concentration of green pebbles some of which are sunk into hollows in the surface. Many of these green pebbles are recognisable as completely glauconitised small *Echinocorys* in which even the test has been pseudomorphed. These echinoids are rather soft as a result of the alteration, and very poorly preserved, in some cases being little more than pebbles approximating to the shape of an echinoid. The matrix of the hardground itself yields numerous elongate *Belemnitella* with less frequent *Cretirhynchia woodwardi* and *Neoliothyrina obesa* Sahni; an aragonite-preservation fauna has not been noted, although rare and poorly preserved pachydiscid ammonites occur. The North Antrim Hardground can be well seen on the shore platform at Portaneevey [D 063 444], in the Larry Bane Head Quarry and in the cliffs at White Rocks and Sliddery Cove. Collecting is best in the quarry, where large blocks litter the floor after blasting.

Supplementary records include:
Porosphaera globularis
Rhizopoterion cribrosum [North Antrim Hardground]
Cyclabacia stellifera Bölsche
Meliceritites undata d'Orbigny
Volviflustrellaria taverensis Brydone
Ditrupa sp.
Terebratulina chrysalis
Chlamys cretosa
Pycnodonte sp.
Spondylus sp.

The Glenarm Chalk marks a second level (the first being the Larry Bane Chalk) of relative abundance of *Cretirhynchia woodwardi* and *Cardiotaxis ex gr. heberti*, both of which terminate at the North Antrim Hardground and are particularly characteristic of the lower part of the *mucronata* Zone in northern Europe. The fauna is closely comparable with that of the upper third of the Weybourne Chalk of the Peake and Hancock (1961) classification of the Norfolk *mucronata* Zone.

---

1 See footnote 2 on p. 85.

Post-North Antrim Hardground succession

The post-North Antrim Hardground succession has already been considered in outline by Wood (*in* Wilson, 1972) who introduced an informal classification into richly fossiliferous *Cardiaster* beds intercalated between two belts of giant flints, and followed by beds characterised by a progressive appearance of Maastrichtian faunal elements. Although these broad lithological divisions have now been superseded, it is nevertheless very helpful in the field to consider the post-North Antrim Hardground succession in terms of this classification, since it provides a framework which can be recognised to a greater or lesser degree in other European *mucronata* Zone successions, particularly the Norfolk Chalk.

The post-North Antrim Hardground succession can be examined conveniently in the Ballymagarry Quarry up to a level which must be assigned to the Maastrichtian on the basis of the occurrence of *Belemnella*. The highest part of the Ballymagarry succession overlaps with the Maastrichtian of the Ballycastle area in Sheet 8 (see Wood, 1967), but detailed correlation between the two areas remains uncertain due to poor exposure and relative inaccessibility of this part of the succession in Ballymagarry Quarry.

The fauna of the *Glenarm β* subdivision (1.57 m at White Rocks) has been collected from Portaneevy, Larry Bane Quarry and the Sliddery Cove–White Rocks cliff section east of Portrush; outside the sheet boundaries there are good collecting conditions in the shore platform of Altachuile Bay on Rathlin Island (Sheet 8). In marked contrast to Glenarm α, the larger fossils are relatively uncommon, although air-weathering reveals an abundance of small fossils, notably bryozoa. In this respect Glenarm α displays similarities with Ballintoy B. Glenarm β marks the entry of the higher *mucronata* Zone fauna characterised by *Cretirhynchia arcuata* and *Cardiaster cordiformis*, two species which attain their acme much higher in the succession. A single example of *Cretirhynchia woodwardi*, which is one of the characteristic species of the lower *mucronata* Zone fauna, was collected from a level 15 cm above the North Antrim Hardground at White Rocks. There is no doubt regarding the horizon, and it remains uncertain whether this anomalous record is to be regarded as a derived fossil from Glenarm α but incorporated in Glenarm β sediment, or whether in fact *C. woodwardi* ranges across the North Antrim Hardground. In the correlative Norfolk succession *C. woodwardi* cuts out immediately below the top of the main Catton Sponge Bed (C. J. W., unpublished). The *Echinocorys* fauna is poorly known, but includes the narrow-based form recorded from Glenarm α, and a globose, relatively thin-tested form. The belemnites that have been collected are badly preserved, and no comment can be made on their affinities. The remaining fauna includes:

Porosphaera globularis
'*Membranipora*' *sp.* indet.
Pavolunulites scandens
?*Reptomulticlausa* cf. *variabilis* (d'Orbigny)
Ubaghsia cf. *gasteri* Larwood
Volviflustrellaria taverensis [Garron Point]
Proliserpula ampullacea
Stereocidaris '*serrifera*': radioles
'*Terebratulina*' *sp.*: i.e. undetermined microbrachiopod
Limea granulata
pectinid, not determined
Spondylus dutempleanus

Garron Chalk   (9.24 m at White Rocks)

The Garron Chalk has its reference section in the cliff south of Garron Point in the east coast area (Sheet 14), where a tripartite subdivision has been established. These subdivisions can be broadly recognised in Sheet 7, but for the purpose of the present contribution Garron Chalk is regarded as undivided. Particularly in the east coast, but perhaps to a lesser extent in the north, the Garron Chalk is characterised by giant flints—many of which are ring flints comparable with those known from the Beeston Chalk in Norfolk. The enormous flint rolls in the foreshore exposures just south of Portantonnish on Rathlin Island referred to by Wilson and Robbie (1966, p. 170) presumably belong here, although their large size (up to 4.5 m in length) is somewhat unusual.

Both in the type area on the east coast and in the present sheet, Garron Chalk occupies a relatively inaccessible level in the cliff sections, and is consequently the least well known subdivision of the post-North Antrim Hardground succession. Information on the echinoid and belemnite faunas is almost entirely lacking as both these groups are poorly represented except in the higher beds of the unit. Weathered sections reveal an abundant mesofauna, and in this respect the Garron Chalk is very similar to the underlying Glenarm β. The whole succession can be examined in a tectonic window (see Figure 20; Plate 32) at the back of the fisherman's path (Larry Bane Rodden) leading from the Larry Bane Quarry down into Larry Bane Bay. The higher beds are exposed between sea level and the first platform in the area known as Portnool [C 897 410] (Plate 36) immediately to the west of The Riggin, between Ballymagarry Quarry and Dunluce, but collecting conditions are rather unfavourable. Collecting by means of a ladder at White Rocks at the point where the North Antrim Hardground descends to beach level would probably yield a good fauna, but this has not yet been attempted.

The mesofauna of the Garron Chalk is particularly rich in bryozoa, to the extent that this part of the succession could be described as a bryozoan chalk. Disarticulated ophiuroids are also very much in evidence. The lists given below are taken largely from the Larry Bane Rodden locality, with additional records from White Rocks. A distinctive flintless bed in the lower part of the succession can be traced from the Garron Point reference section into Sheet 7, and the rich faunas obtained from this level have been included here for the sake of completeness: Garron Point records are designated by an asterisk.

A: LOWEST BEDS
Porosphaera globularis
Clausa globulosa (d'Orbigny)
Lunulites patelliformis Marsson
Onychocella pyriformis (Goldfuss)
Thyracella postifera Voigt
Volviflustrellaria taverensis
Terebratulina chrysalis: juv.
asteroid ossicles, including *Ophryaster sp.* and *Pycinaster sp.*
Glomerula gordialis
Neomicrorbis sp.: *N.* (*Granorbis*) *salebrosus* Regenhardt?
Inoceramus shell fragments
Neithea sexcostata
Belemnitella sp. ex gr. *langei* Jeletzky?

B: FLINTLESS BED AND IMMEDIATELY ADJACENT CHALK
*Coelosmilia wiltshirei Duncan
*Beisselina cf. *pachyphylla* Voigt
*Castanopora castanea Lang
*Clausa variabilis (Hagenow)
*Lichenopora stellata (Goldfuss)
*Lunulites beisseli Marsson
L. cf. *mitra* Hagenow
L. semilunaris (Hagenow)
**melicerititid, not determined
*'Membranipora' cf. *cubitalis* Brydone
*Nodelea semiluna (d'Orbigny)
*Onychocella dichotoma (Marsson)
O. matroma? (Hagenow)

Pavolunulites sp. nov.
Patalopora reticulata (Marsson)
Spiropora cf. *macropora* (d'Orbigny) *micropora* Gregory
'Vincularia' cf. *strumulosa* (Marsson)
Volviflustrellaria taverensis
Ditrupa (Pentaditrupa) subtorquata (Münster)
Glomerula gordialis
Neomicrorbis crenatostriatus (Münster)
N. (Granorbis) salebrosus
Cretirhynchia norvicensis
Terebratulina chrysalis
Ophiomusium granulosum
Ophiura
*cidaris radioles; including 'serrifera' Auctt.
Galerites cf. *vulgaris*
Phymosoma sp.
Salenia sp.
Belemnitella sp. ex gr. *langei*?

C: HIGHER BEDS, MAINLY FROM LARRY BANE RODDEN
Porosphaera globularis
Entalophora cf. *echinata* (Roemer)
Homoeosolen ramulosus Lonsdale
Meliceritites semiclausa (Michelin)
M. sp. nov.
'*Membranipora*' cf. *acuum* Brydone
'*M.*' sp. nov.: ex gr. *M. dunensis* Brydone
Onychocella cf. *dichotoma*
Pavolunulites richteri and *P.* cf. *richteri*
P. scandens Brydone
pelmatoporine cribrimorph, not identified
Volviflustrellaria taverensis
Carneithyris sp.
Cretirhynchia arcuata
C. norvicensis
Terebratulina chrysalis
Ophiomusium granulosum
indeterminate small echinoid plates and radioles: cidarids, saleniids
Limea granulata
Belemnitella spp.: including *B.* cf. *langei*
B. sp. = giant form figured by Jeletzky (1964, pl. 1, fig. 4 as *Belemnitella mucronata*)

PORTRUSH CHALK (14.44 m at Ballymagarry)

This unit, the type section of which is taken in the Ballymagarry Quarry 3 km east of Portrush (Plate 37), corresponds to the 'Cardiaster Beds' of Wood (*in* Wilson, 1972), and is relatively thin-bedded in contrast to the more massive chalk below. Concentrations of fragmented *Inoceramus* shell are present at several levels, and the general picture is one of disturbed and probably relatively shallow-water sedimentation. The Portrush Chalk is richly fossiliferous, with a diverse benthonic fauna dominated by brachiopods and large echinoids; compared with the Garron Chalk, however, bryozoa are little in evidence.

The Portrush Chalk is subdivided into four beds, coded A to D in ascending order. A noticeably marly plane—possibly an actual marl in an unweathered section—occurs at the top of the third bed (C), and forms a useful field marker. The equivalent horizon in the Port More Borehole is expressed as a minor negative deflection in the resistivity profile.

Portrush A is characterised by tabulate and lensoid flints which tend to develop along bedding-planes, while the remainder of the unit exhibits the more normal burrow-fill type of flint. The paucity of fossils in the lower part of Portrush A contrasts strongly with the richness at higher levels. Intensive collecting from the Portnool Platform section yielded infrequent small *Cretirhynchia* ex gr. *norvicensis* Pettitt, slim and very elongate small *Belemnitella* (?*langei* group) and a depressed flat-based *Echinocorys* with a trapezoidal profile. A globose thin-tested form was also collected. The typical Portrush fauna characterised by *Echinocorys* cf. *conoidea* (Smiser, 1935, fig. 18a–c) and *E.* aff. *conoidea* (see Peake and Hancock, 1961, p. 318) sets in in the topmost metre of Portrush A where there is an *Echinocorys* bed. The latter form is diagnostic of the Beeston Chalk subdivision of the Norfolk *mucronata* Zone, in which it locally dominates the *Echinocorys* assemblage. The Portrush brachiopod fauna also compares well with that of the Beeston Chalk: *Carneithyris* becomes common for the first time, notably in the top 10 cm of Portrush C, where it is associated with *Cretirhynchia arcuata*. The occurrence of *Orbirhynchia* together with the rare kingenid *Kingenella* matches recently reported finds in the Beeston Chalk of the Norwich area (Wood, 1972), where the benthonic fauna attains its maximum diversity. The belemnite fauna is dominated by *langei* group forms, with subordinate giant *Belemnitella mucronata* sensu Jeletzky (Jeletzky, 1964, pl. 1, figs. 4a–c; Christensen and others, 1973) which are virtually restricted to this horizon. *Galerites roemeri* (d'Orbigny), originally described from the equivalent of the Beeston Chalk in Germany (see d'Orbigny, 1853–60, pl. 1006, figs. 1–6) is another characteristic Beeston faunal element found at this level.

The fauna of Portrush D (i.e. above the marly plane) differs somewhat from that of beds A–C. The diagnostic *Echinocorys* cf. *conoidea* and *E.* aff. *conoidea* continue from below, but appear to die out before the top of the bed. A medium-sized globose/pyramidate form of uncertain affinities, which ranges throughout the succeeding Chalk, enters in the upper part of Bed D, where it is associated with abundant small *Galerites*. The diverse Portrush brachiopod fauna is also well represented, and a large rhynchonellid provisionally referred to *Cretirhynchia* aff. *arcuata* is particularly characteristic. Belemnites are much reduced in number and confined to *Belemnitella* of the *langei* group.

Wood (*in* Wilson, 1972) stated that the unit now designated Portrush Chalk was characterised by *Cardiaster granulosus* (Goldfuss). Ernst (1972), however, drew attention to the fact that the higher *mucronata* Zone (i.e. Beeston Chalk) *Cardiaster* should be referred to *C.* aff. *granulosus*, since it differed in plastronal structure from the true *C. granulosus* which is Maastrichtian. It is this Beeston form, superbly figured by Forbes (1852, pl. 9), which occurs in the Portrush Chalk, and for which it would seem appropriate to use the name *cordiformis* introduced by Woodward (1833). Very few specimens of *C. cordiformis* have been collected *in situ*, although a number of loose specimens undoubtedly belong here. In the south-west of the Cretaceous outcrop, however, the equivalent beds are present in a condensed sequence at the base of the Chalk succession, e.g. at Carmean near Moneymore, and here *C. cordiformis* is strongly represented, together with other distinctive Portrush/Beeston elements such as *Galerites roemeri* and *Echinocorys* aff. *conoidea*. An unlocalised flint steinkern of *C. cordiformis*—presumably from the Clay-with-Flints in Co. Londonderry—was figured by Portlock (1843, pl. 17, figs. 2 a–c) as *Holaster aequalis*.

Records supplementary to those of the range charts are as follows:
Portrush A:
  Porosphaera sp.
  Pavolunulites scandens
Portrush B:
  Porosphaera sp. [pyriform]
  bryozoa, including *Onychocella sp.* and *Volviflustrellaria sp.*
Portrush C:
calcisponge: *Porosphaera*?
  octocoral?
  '*Membranipora*' cf. *cubitalis*
  '*M.*' sp. nov. [*acuum* group]

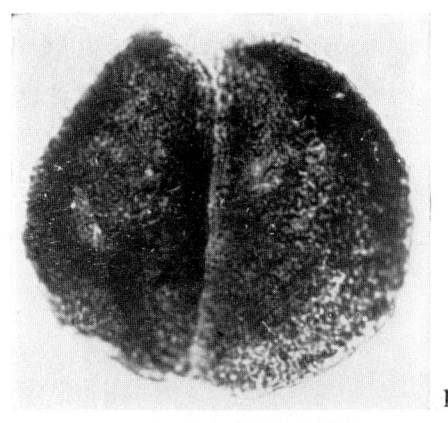
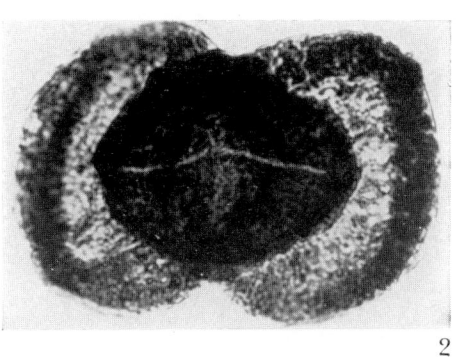
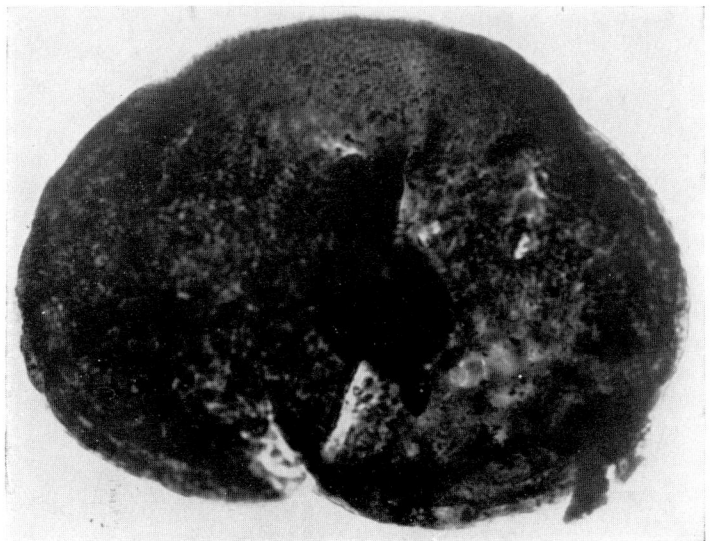
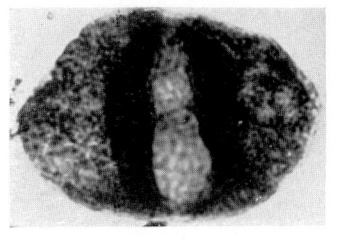
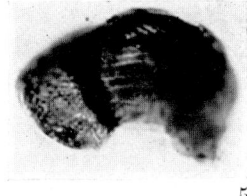
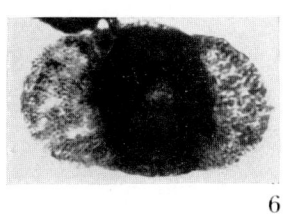
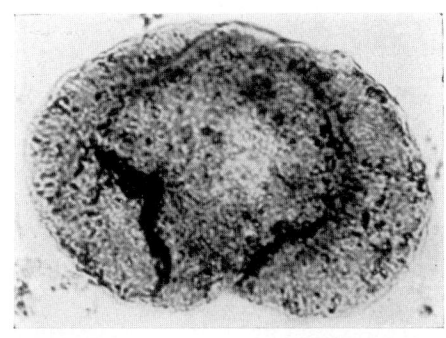
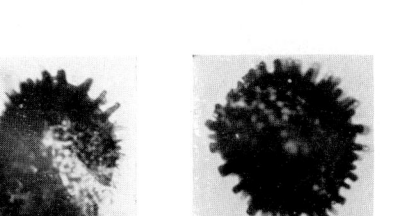
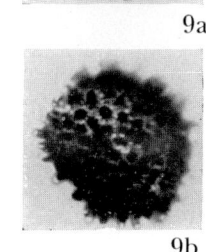

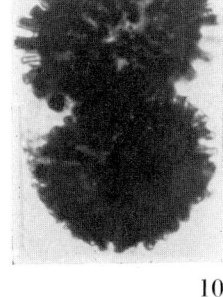
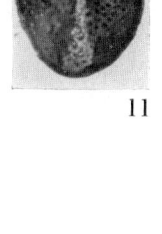
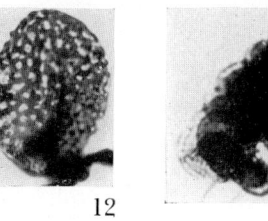
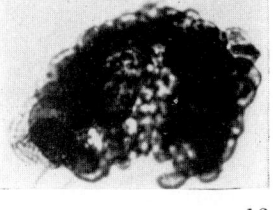

Plate 24
Triassic miospores from the Glenstaghey Formation in the Port More Borehole

1 *Sulcatisporites kraeuseli* Mädler. (MPK. 624)
2 *Angustisulcites klausii* Freudenthal 1964. (MPK. 625)
3 *Brachysaccus neomundanus* (Leschik) Mädler 1964. (MPK. 626)
4 *Alisporites grauvogeli* Klaus 1964. (MPK. 627)
5 *?Striatoabieites aytugii* Visscher emend. Scheuring 1970. (MPK. 628)
6 *Triadispora sp.* (MPK. 629)
7 cf. *Colpectopollis ellipsoideus* Visscher 1966. (MPK. 630)
8 *Echinitosporites iliacoides* Schulz & Krutzsh 1961. (MPK. 631)
9a *E. iliacoides* (median focus). (MPK. 632)
9b *E. iliacoides* (high focus). (MPK. 632)
10 *E. iliacoides*. (MPK. 633)
11 *Retisulcites perforatus* (Mädler) Scheuring 1970. (MPK. 634)
12 *R. perforatus*. (MPK. 635)
13 *Tsugaepollenites oriens* Klaus 1964. (MPK. 636)

All specimens are illustrated at a magnification of ×600 and are held in the Institute of Geological Sciences palynology collection at Leeds. The figured specimens are located within linked rings on the slides and are registered in the MPK catalogue.

102 CHAPTER 15 CRETACEOUS ROCKS

Plate 25 The Giant's Cut. About 75 m of Lower Basalts, including a 9-m bed of tuff, overlie the White Limestone

(*Photograph E. Fairclough*)

Plate 26 The basal conglomerate (Hibernian Greensands) and the lowest three members of the White Limestone at Oweynamuck
Hib. Gr. = Hibernian Greensands; Gal. = Galboly Chalk; Clog. = Cloghastucan Chalk; OFB = Oweynamuck Flint Band; Creg. = Creggan Chalk.

(*Photograph R. E. H. Reid*)

Plate 27 Cliffs of Boheeshane Chalk with Creggan Chalk at the west end of White Park Bay
Cre. = Creggan Chalk; BPB = Bendoo Pebble Bed; Boh. = Boheeshane Chalk

Plate 28 Cliffs of Boheeshane to Ballintoy Chalk members of the White Limestone at Portbraddan, with the *Belemnitella praecursor* Bed exposed on the rock platform
Boh. = Boheeshane Chalk; Orb. Band = *Orbirhynchia* Band; Lar. = Larry Bane Chalk; Bal. = Ballintoy Chalk

(*Photographs R. E. H. Reid*)

104 CHAPTER 15 CRETACEOUS ROCKS

Plate 29
Eastern spring section in the Boheeshane Chalk at the western end of White Park Bay
Cre = Creggan Chalk; BPB = Bendoo Pebble Bed;
Boh. = Boheeshane Chalk; Orb. Band = *Orbirhynchia* Band

(*Photograph R. E. H. Reid*)

Plate 30
Cliff section at Larry Bane Head, west side, showing the White Limestone succession from the Cloghastucan Chalk to the Larry Bane Chalk
Clo. = Cloghastucan Chalk; OFB = Oweynamuck Flint Band;
Cre. = Creggan Chalk; BPB = Bendoo Pebble Bed;
Boh. = Boheeshane Chalk; Lar. = Larry Bane Chalk

(*Photograph R. E. H. Reid*)

Plate 31 Aerial view of cliffs of Creggan to Ballintoy Chalk members of the White Limestone at the western end of White Park Bay
Cre. = Creggan Chalk; Boh. = Boheeshane Chalk; Lar. = Larry Bane Chalk; Bal. = Ballintoy Chalk

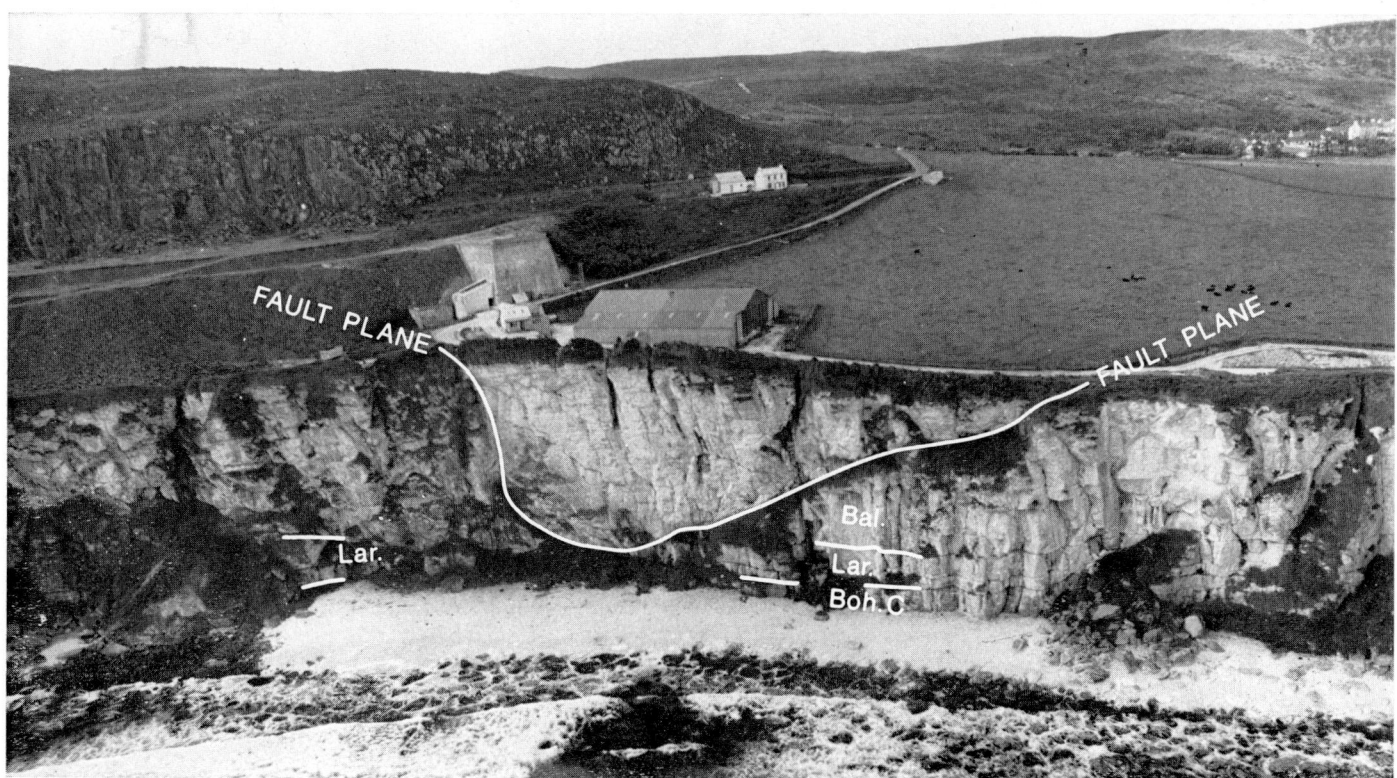

Plate 32 Cliffs of Boheeshane to Ballintoy Chalk on the east side of Larry Bane Head. The tectonic window, inland of the fault plane indicated, shows high members of the White Limestone above the North Antrim Hardground
Boh. = Boheeshane Chalk; Lar. = Larry Bane Chalk; Bal. = Ballintoy Chalk

(*Photographs E. Fairclough*)

Plate 33 Cliffs west of Larry Bane Head showing the Boheeshane to Garron Chalk members of the White Limestone
Boh. = Boheeshane Chalk; OB = *Orbirhynchia* Band; WFB = Whitehead Flint Band;
Lar. = Larry Bane Chalk; Bal. = Ballintoy Chalk; Gle. = Glenarm Chalk; NAHG = North Antrim Hardground;
Gar. = Garron Chalk

(*Photograph R. E. H. Reid*)

Plate 34 Cliffs of White Limestone at Larry Bane Head and Quarry
Boh. = Boheeshane Chalk; Lar. = Larry Bane Chalk; Bal. = Ballintoy Chalk; Gle. = Glenarm Chalk; Gar. = Garron Chalk

Plate 35 Cliffs of White Limestone at Boheeshane Bay. Ballintoy village lies at the foot of the basalt scarp in the background
Boh. = Boheeshane Chalk; Lar. = Larry Bane Chalk; Bal. = Ballintoy Chalk

(*Photographs E. Fairclough*)

103 CHAPTER 15 CRETACEOUS ROCKS

Plate 36 Cliffs from The Riggin to Jackstone Cave. The Portrush Chalk Member is well exposed on the Portnool platform

Plate 37 White Rocks and Ballymagarry Quarry. Glenarm to Portrush Chalk members in the sea cliff overlain by Ballymagarry and higher members in the quarry
NAHG = North Antrim Hardground; By = Ballymagarry Chalk; LGFB = Long Gilbert Flint Band

(*Photographs E. Fairclough*)

Metopaster tumidus: marginals
Neithea sexcostata
Spondylus dutempleanus
Portrush D:
Porosphaera sp.
'*Membranipora*' sp. nov. [*acuum* group]
Pelmatopora sp.
Cretirhynchia ex gr. *norvicensis* [small form]
asteroid marginals, including *Metopaster tumidus* and
Pycinaster angustatus?
Pseudoptera coerulescens

BALLYMAGARRY CHALK   (10.6 m at Ballymagarry)

The Ballymagarry Chalk, which takes its name from the type section in the Ballymagarry Quarry, equates with the 'upper belt of giant flints' (Wood *in* Wilson, 1972, fig. 22), and is the approximate correlative of the Paramoudra Chalk of the Norfolk *mucronata* Zone. The member is subdivided into three beds coded A to C, and its upper limit is taken at the base of the Long Gilbert Flint Band, which is provisionally regarded here as a close approximation to the position of the Campanian–Maastrichtian boundary. The Ballymagarry Chalk would thus represent the uppermost Campanian—a part of the succession which is very poorly exposed in Norfolk.

Ballymagarry A is marked by a continuation of the abundance of small *Galerites* observed in the upper part of Portrush D, although here the diverse brachiopod fauna is absent, the only brachiopod recorded being a small *Cretirhynchia* of the *norvicensis* group. The complete absence of terebratulids (*Carneithyris*) compared with their abundance in the beds above and below is particularly striking. A single specimen of *Cardiaster cordiformis* was collected from a low position in the bed. Higher levels have yielded the earliest *Echinocorys belgica* Lambert, a relatively thick-tested essentially Maastrichtian species, which ranges throughout the Ballymagarry Chalk, but which is particularly characteristic of the overlying Tanderagee Chalk. The only other records are an indeterminate *Phymosoma* radiole and serpulids referred to *Ditrupa* (*Tetraditrupa*) *sp*.

Ballymagarry B is characterised by massive chalk with irregularly spaced giant flints, some of which are true paramoudras. In the Ballymagarry section paramoudras are rare, although in correlative sections in the Belfast area (Manning and others, 1970; Wood *in* Wilson, 1972) they represent the characteristic type of flint development. Ballymagarry B (and the overlying Bed C) are well exposed in the south wall of the main Ballymagarry quarry, but access to them can be obtained only from the steeply sloping talus to one side of this section. In consequence only limited faunas have been collected from these beds, with the exception of the lowest metre, and these are almost certain to be unrepresentative. Ballymagarry B is thought to correlate with the lowest massive flinty beds exposed in the shore platform beyond the first deep inlet to the west of the Port More Fault (see Figure 21), although it should be noted that at this point true ring flints are developed.

The Ballymagarry B fauna is more diverse than that of Ballymagarry A, and marks a return of *Carneithyris* and occasional belemnites of the *langei* group. Small *Galerites* are common in the lower beds, which have also yielded *Ditrupa* (*Tetraditrupa*) and masses of *Castonopora magnifica* (d'Orbigny) preserved in flint. The few poorly preserved *Echinocorys* so far collected include the globose pyramidate forms with round base, and massive depressed forms which, by analogy with Norfolk, are probably comparable with the forms characterising the chalk of the Thorpe Asylum pit [Rowe Collection, British Museum (Natural History)].

Ballymagarry C continues the sequence of massive flints, but the higher beds are comparatively flintless and include several well developed bedding-planes, some of which may represent omission surfaces. Collecting is limited by the relatively inaccessible position of this bed in the quarry face. Ballymagarry C overlaps with part of the succession exposed in the Port More platform (see Figure 21), but it has not proved possible to correlate the two sections in detail. Fletcher (*in press*) suggested that the master bedding-plane in the Port More section corresponds to the Long Gilbert Flint at the base of the Tanderagee Chalk type-section. The Port More succession below the master bedding-plane includes a flintless belt yielding *Belemnitella* ex gr. *langei*, a medium-sized *Echinocorys* with an elongate base and an asymmetrical lateral profile, and a small

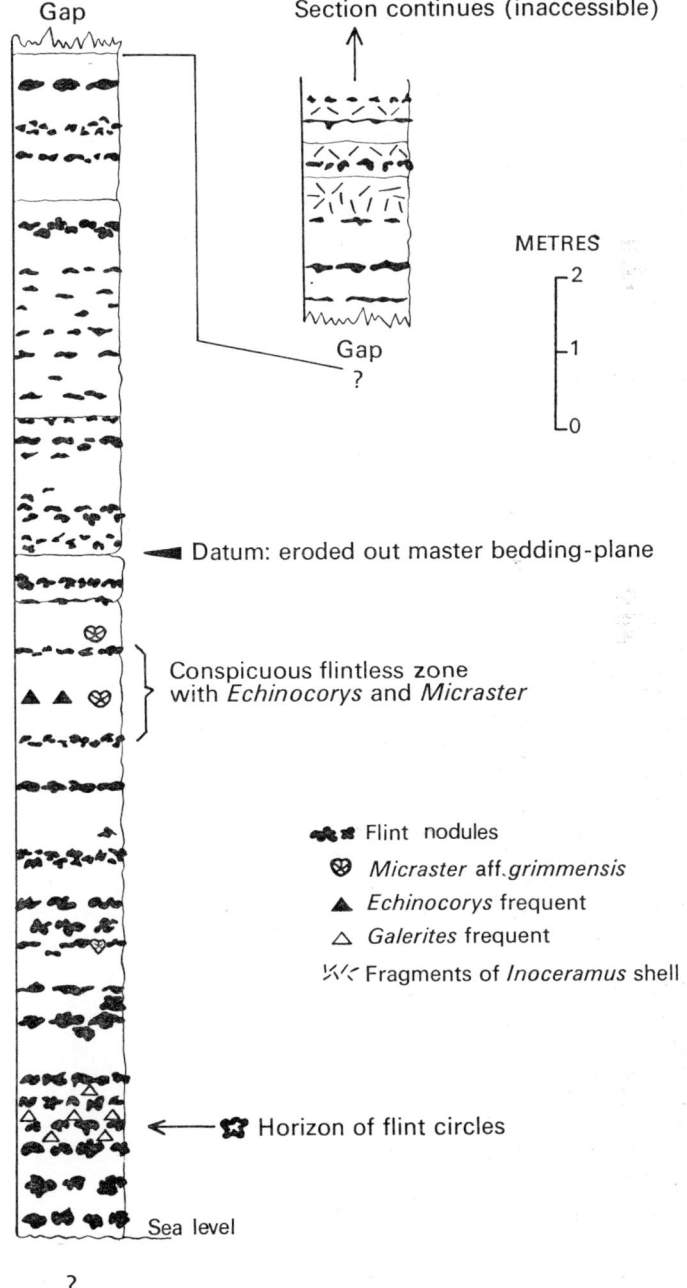

Figure 21   Port More: composite cliff and platform succession

Micraster. The same *Micraster* was recorded from the presumed correlative 'Top Flinty Beds' of the Belfast area (Wood *in* Manning and others, 1970) as *Micraster (Isomicraster) sp.*, and is well represented at various levels in the upper part of the Paramoudra Chalk of Norfolk, notably in the Rowe collection [BM (NH)] from the Thorpe Asylum pit. A block of flint with several decalcified examples which was almost certainly collected from the Clay-with-Flints is registered in the Ulster Museum collections under the accession number K 2780. M. G. Schulz, Hamburg (unpublished), who has collected about 200 specimens of *Micraster* from the Campanian–Maastrichtian boundary succession exposed at Kronsmoor near Lägerdorf in Holstein (West Germany), has demonstrated that the Thorpe *Micraster* is closely related to *M. grimmensis* Nietsch (Nietsch, 1921, pl. 10, figs. 9a–c), but differs in that the anterior profile is distinctly concave. [This form may be *M. ciplyensis* Schlüter, a poorly known species which occurs in the Craie de Spiennes and (perhaps as a remanié fossil) in the overlying Craie de Ciply: see Schlüter, 1897, pp. 18–19, pl. 2, figs. 1, 2; Lambert, 1897, pp. 45–46, pl. 2, figs. 1, 2; Lambert, 1911, pp. 43–44, pl. 2, fig. 16; Smiser, 1935 (sub *Isomicraster ciplyensis*), p. 83]. At Kronsmoor *M.* aff. *grimmensis* occurs as a rarity together with *M. grimmensis* in the uppermost Upper Campanian. The *stratum typicum* of *M. grimmensis* at Grimme (East Germany) is probably the same horizon. In Britain, true *M. grimmensis* is associated with *M.* aff. *grimmensis* in the derived assemblage in the Cromer Lighthouse flint mass, Norfolk (Peake and Hancock, 1970, pp. 339F–G)—Schulz, personal communication—and a single specimen is in the Brydone collection (Norwich Castle Museum no. 76.937) from an unspecified chalk mass at Overstrand. At present it is not known whether true *M. grimmensis* occurs above *M.* aff. *grimmensis* or whether *M. grimmensis* is an eastern geographical subspecies. If the former, it is possible that true *M. grimmensis* in Norfolk may come from masses which can be shown on belemnite evidence to be at least in part of Maastrichtian age. All that can be stated here regarding the Northern Ireland succession is that there is a frequency-maximum of the *M. grimmensis* group in the presumed correlative of the Ballymagarry Chalk at Port More, and that the group has not yet been recorded from the overlying Tanderagee Chalk. The Ballymagarry Chalk is accordingly assigned to a very high Upper Campanian position.

The limited fauna from the Ballymagarry C type section includes the essentially Maastrichtian large rhynchonellid *Cretirhynchia magna* Pettit, and a small pyramidate *Echinocorys – E. pyramidata* Portlock sensu Wright (unsymmetrical form) (Wright, 1864–82, pl. 77, fig. 3) which occurs in the derived blocks of hard chalk in the flint masses west of Overstrand (Peake and Hancock, 1970, p. 339G). This hard chalk is thought by Peake and Hancock to represent a terminal Campanian hardground, but the relation between this and the Sidestrand masses about 1 km to the east has not been satisfactorily established. The Ballymagarry C fauna is clearly transitional between that of the high Campanian and Maastrichtian, although the occurrence of *Belemnitella* of the *langei* group indicates that it should be placed in the Campanian. *Austinocrinus bicoronatus* (Hagenow), which elsewhere ranges across the Campanian–Maastrichtian boundary, has not yet been recorded. Supplementary fauna from Ballymagarry C includes: *Porosphaera sp.* (pointed form), *P. globularis*, *Ditrupa sp.*, *Pavolunulites scandens*, asteroid marginals including *Metopaster tumidus*, *Terebratulina chrysalis* and *Spondylus dutempleanus*.

TANDERAGEE CHALK (7.64 m at Ballymagarry)

The Tanderagee Chalk, named after the townland adjacent to the Ballymagarry quarry where the type section is taken, is delimited below by the Long Gilbert Flint, and terminates at a bedding plane 30 cm below a conspicuous complex tabular flint with downward prolongations resembling drops of liquid—referred to subsequently as the 'drip-flint'. The Tanderagee fauna is characterised by the crinoid *Austinocrinus bicoronatus*, which ranges in Norfolk from the Beeston Chalk to the pre-*Porosphaera* Beds and *Porosphaera* Beds subdivisions of the Maastrichtian (Wood, 1967; Peake and Hancock, 1971), being particularly well represented in the Maastrichtian part of the succession. The belemnites are exclusively *Belemnitella*, but the *langei* group forms of the underlying members are replaced abruptly by larger belemnites provisionally referred to *B. posterior* Kongiel (Kongiel, 1962, pl. 19, figs. 1–3) which can be recognised by their conical lateral profile and strong vascularisation. In the Kronsmoor section *Belemnitella* of the *langei* groups cuts out before the Campanian–Maastrichtian boundary, while other *Belemnitella* groups continue (Schulz, unpublished); it is therefore particularly significant that this group has not been recorded above the top of the Ballymagarry Chalk. Reassessment of the belemnites from the lower two zones of the informally named Ballycastle Chalk (Wood, 1967) and its Norfolk correlatives has shown that the identification of *Belemnitella langei* was erroneous: the *Belemnitella* assemblage comprises predominantly various growth stages of *B. posterior*, together with small pseudogranulate cylindrical forms which may be early members of the essentially Upper Maastrichtian group of *B. junior* Nowak. The absence of the diagnostic Maastrichtian genus *Belemnella* from the Tanderagee Chalk does not preclude these beds from being of Maastrichtian age, since it is now known (Schultz, unpublished) that *Belemnella* is extremely rare at the base of the stage in the Kronsmoor section, which appears to be the most complete biostratigraphical record available of the critical Campanian–Maastrichtian boundary succession. The *Belemnitella posterior/Austinocrinus bicoronatus* association which characterises the Tanderagee Chalk is, however, well represented in the lower part of the pre-*Porosphaera* Beds of the Overstrand Hotel Lower Mass, Norfolk (Wood, 1967) only 2 to 3 m below the occurrence of a mixed *Belemnella/Belemnitella* assemblage in which the *Belemnella* are considered by Schulz (personal communication) to be already relatively advanced forms. Other fossils common to both the lower pre-*Porosphaera* Beds and the Tanderagee successions are thick-tested depressed and/or subpyramidal *Echinocorys* of the *belgica* group (Lambert, 1897, pl. 4, figs. 9–10), and a small pisiform undescribed rhynchonellid referred to here as *Cretirhynchia sp. nov. A*.

The interval between the Long Gilbert Flint and the next flint course yields numerous *Echinocorys* ex gr. *belgica* and *Belemnitella posterior*; large gryphaeate *Pycnodonte* also occur. A second level with *E.* ex gr. *belgica* is found between two distinctive paired flint courses. The chalk immediately below the lower of the paired flint courses is rich in echinoderm debris and other small fossils, of which the following can be recorded:

Porosphaera sp. [small, discoidal]
P. globularis
Cyclabacia clathrata (Hagenow)
Lateroflustrellaria hexagona d'Orbigny
Glomerula gordialis
Ophiomusium granulosum
Austinocrinus bicoronatus
asteroid marginals including *Pycinaster* cf. *crassus*, *Recurvaster?*
Stereocidaris 'serrifera': radioles
Phymosoma sp. juv. cf. *magnificum*

Cretirhynchia magna was recorded from the lower paired flint. From above the upper paired flint a fauna of small brachiopods was collected, including *Cretirhynchia sp. nov. A*, *Orbirhynchia sp.* and *Carneithyris sp.*, which compares with that of the Norfolk pre-*Porosphaera* Beds. With the exception of the lowest metre, belemnites are uncommon in the lower part of the Tanderagee Chalk, and do not reappear in strength until the upper levels: the higher belemnites are all small *Belemnitella sp.*, but no determinable specimens have been collected. The highest beds again yield a fauna of small fossils, including, in addition to the diagnostic *Austinocrinus bicoronatus*, the following species:

Porosphaera globularis
P. nuciformis (Hagenow)
Pavolunulites scandens
Metopaster sp.: marginal
Spondylus dutempleanus

POST-TANDERAGEE CHALK SUCCESSION (Units designated numerically; 9.88 m measured plus an estimated 3.65 m to the contact with the Lower Basalts)

This part of the succession is accessible only in the western extension of the main Ballymagarry quarry, where it is exposed in a number of fault-bounded blocks. Poor exposure, coupled with the effects of numerous minor displacements, has made these sections particularly difficult to measure and correlate, and it is possible that the composite succession given here may contain some inaccuracies. Difficulties of correlation of isolated fault blocks has limited the collecting of faunas on a bed-by-bed basis, and only the air-weathered faces of Unit 1 have been studied in detail. The Maastrichtian age of the top beds of Units 3 and 4 respectively is established by the occurrence of *Belemnella lanceolata*. *Belemnitella posterior* occurs sparingly in Unit 1, but no examples of *Belemnella* have been recorded so far. Unit 1 has also yielded *Echinocorys pyramidata* Portlock (sensu Wright *non* Lambert), small *Galerites sp.*, *Pycnodonte vesicularis* and a thin-shelled wide form of *Cretirhynchia retracta* (Roemer) which characterises the Norfolk *Porosphaera* Beds, and which approaches *C. magna* in external characters. The thin-bedded sediments of Unit 2 contain numerous small brachiopods amongst which the rare kingenid *Kingenella* is noteworthy. This genus was previously noted from the Portrush Chalk (and the correlative Beeston Chalk of Norfolk), and its occurrence at this level provides an interesting parallel with rare records from the *Porosphaera* Beds (IGS collections).

The Post-Tanderagee Chalk overlaps with the succession described by Wood (1967) from the coast between Port Calliagh and Ballycastle in the Ballycastle (8) Sheet. The latter was informally named the Ballycastle White Limestone, but Fletcher (*in press*) has now reclassified this succession on a formal basis into two members. The lower member (Port Calliagh) corresponds with units A to C of the original section (Fletcher *in* Wood, 1967, pl. 21) together with the underlying succession down to the separation-plane below the 'belt of complex flints'; the remainder of the Ballycastle succession is now referred to as the Ballycastle Chalk Member. It is thought that the 'belt of complex flints' just above the base of the Port Calliagh Chalk correlates with the 'drip flint' of Unit 1 of the post-Tanderagee succession in the Portrush area. If this is correct, the post-Tanderagee succession equates with the Port Calliagh Member, and does not extend as high as in Sheet 8. In the original description of the Ballycastle succession, the Campanian–Maastrichtian boundary was drawn at the junction of units A and B, that is D–E of the recoded Port Calliagh succession. This level probably correlates with a position within Unit 2, in which case Unit B of the original succession would correspond with Unit 3. It is now known, however, that *Belemnella* occurs below this level at Ballycastle, although the lower limit of its range has not been established. From the preceding discussion (pp. 109 and 110) it is obvious that the Campanian—Maastrichtian boundary needs to be lowered, but there is no clear indication at which level it should be drawn, since the available palaeontological evidence is equivocal. The level chosen here is more in the nature of a downward limit, since the Ballymagarry Chalk is clearly Campanian; what remains to be decided is the age of the Tanderagee Chalk. Whatever the eventual decision, the correlation postulated here between the upper beds in the Ballymagarry quarry and the Ballycastle succession necessitates considerable modification of the currently accepted limits of the Maastrichtian in Northern Ireland.

The White Limestone of the Port More Borehole

The Port More Borehole proved a thickness of 91.13 m of Chalk between the depths 77.21 m and 168.35 m. The original lithological log was prepared by P. I. Manning, and about 600 palaeontological specimens were collected, together with a number of representative lithological samples. In addition, a variety of geophysical logs were made, which are reproduced at a small scale in Figure 30.

The core was logged before the recent work on the lithostratigraphy of the Causeway Coast Chalk succession, and consequently the key lithological and palaeontological marker horizons were not recognised. T. P. Fletcher has subsequently examined 11 m of core below the depth 157.35 m. Above this level, with the exception of the North Antrim Hardground, the log records little more than the flint courses which were represented in the core and is not sufficiently informative to be included in the memoir. The succession of flint courses, particularly above the North Antrim Hardground, does not match the complete standard succession at Ballymagarry Quarry, or the partial sections measured in the Port More Cliffs close by and in the fisherman's path section near Larry Bane Head.

The palaeontological record is unsatisfactory. Much of the material is of poor quality, comprising predominantly crushed brachiopods, incomplete belemnites and indeterminate fragments of molluscan shell and echinoderm test, together with occasional more or less complete *Echinocorys*. No fossils whatsoever were preserved from the 7.31 m interval between the depths 94.49 m and 101.80 m; and a smaller gap in the records exists between the depths 91.13 m and 92.35 m. The more important palaeontological information is given below in note form following the discussion of the stratigraphy of the borehole succession. Because of the inadequate lithological and fossil records, interpretation of the Port More Chalk succession has been assisted by the supplementary information provided by the Laterolog, reproduced at large scale in Figure 22. Inspection of this shows a positive drift in values upwards from the Lias contact at 168.35 m until values approaching infinity are reached at a depth of about 143.25 m. Three sharp low resistivity zones, which occur at 140.28 m, 137.92 m and 132.66 m, are separated by intervals having a ratio approximating to 1:2; and clearly correspond to the surfaces delimiting the two beds comprising the Larry Bane Member, thus permitting classification of the borehole succession into pre-Larry Bane, Larry Bane and post-Larry Bane sequences. Above this level, values continue at or near infinity until a change in resistivity level is encountered between the depths 113.7 m and 114.66 m. This change corresponds to the North Antrim Hardground complex, the upper (main) pebble bed of which was recorded in the lithological log at 114.68 m. Similar deflections can be observed in the formation density and sonic logs (see Figure 30).

NOTES ON SELECTED ITEMS OF THE STRATIGRAPHICAL PALAEONTOLOGY OF THE CRETACEOUS IN THE PORT MORE BOREHOLE

1 77.85–79.93 m interval: small asymmetrical *Echinocorys* ('Overstrand type'—see p. 110) [NIE 6853]; although somewhat distorted, this specimen is comparable with CJW 2121 from Ballymagarry C at Ballymagarry.
2 79.93–82.08 m interval: *Echinocorys* cf. *belgica* Lambert; this species characterises basal Tanderagee Chalk, but comparable forms are also known from the Ballymagarry Chalk.

112 CHAPTER 15 CRETACEOUS ROCKS

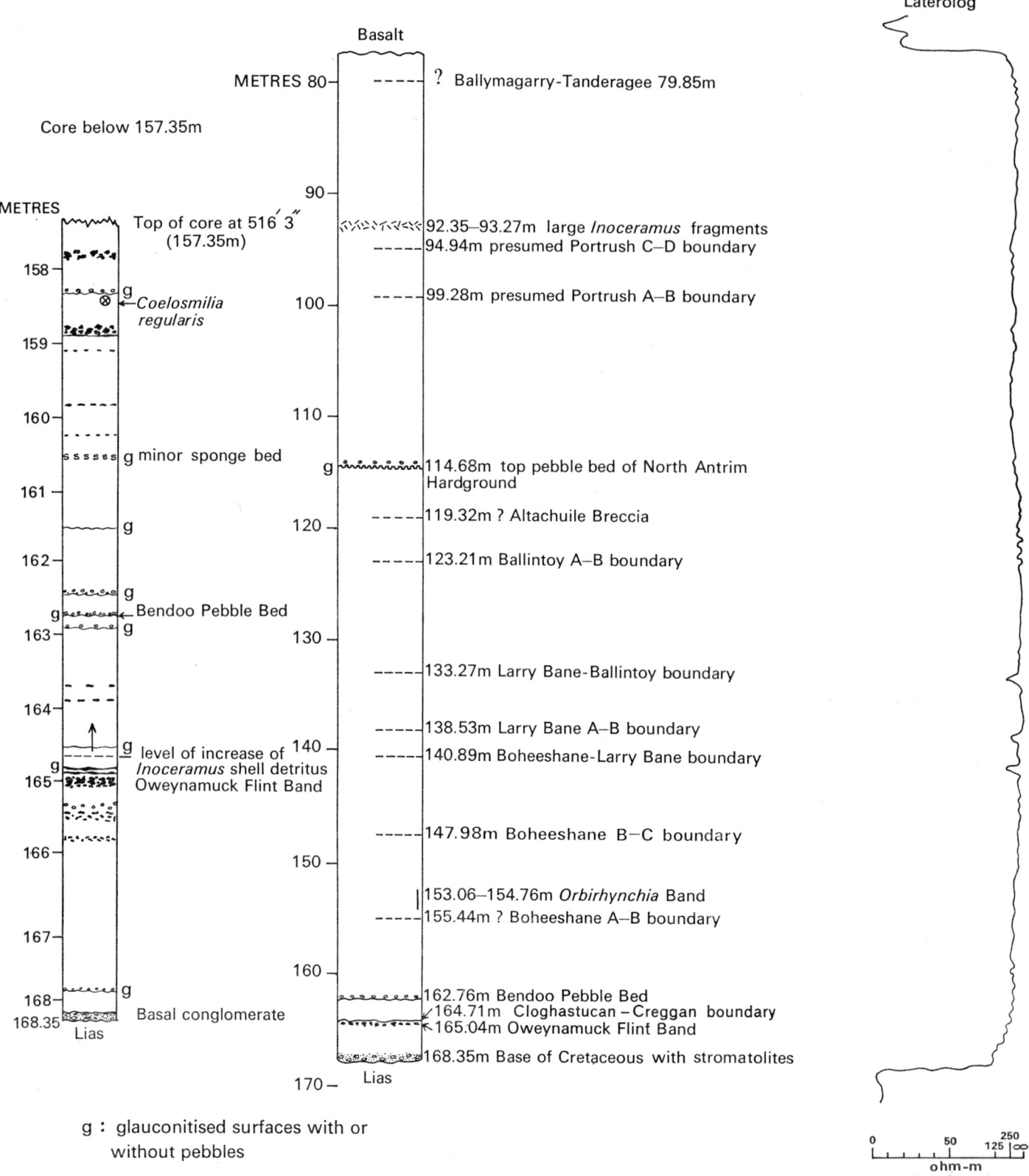

Figure 22 Port More Borehole. Graphic sections of the Cretaceous rocks. The lowest part of the core, logged in more detail, is given on the left

3 A sedimentary surface occurs at 83.56 m; this surface exhibits slight green (glauconite?) staining, and incorporates fragmentary echinoids and corroded belemnites.

4 84.43–86.11 m and 89.92–91.13 m intervals respectively: *Belemnitella* with large fissure-angle—i.e. from the group of *B. langei* Jeletzky, one definite specimen from each interval.

5 92.35–93.27 m interval: horizon of large *Inoceramus* fragments (sheets), suggestive of basal Portrush D.

6 101.80–102.12 m interval (? near base of Portrush Chalk): small *Echinocorys* [NIE 6926] closely comparable with *Echinocorys* ex gr. *conica* (Agassiz) characterising Larry Bane and basal Ballintoy. Although this is an apparently anomalous record, small echinoids of this type are known to occur in the Runton Chalk Erratic on the Norfolk coast—i.e. the equivalent of a post-North Antrim Hardground horizon.

7 112.39–113.84 m: questionable determination of calyx of *Bourgueticrinus* cf. *hagenowi* (Goldfuss) from a horizon attributed to Glenarm β. *B. hagenowi* is recorded from the Lower and Upper Maastrichtian on the Continent and from the Lower Maastrichtian in England. Note that even the determination of this specimen as a crinoid calyx is uncertain.

8 116.43–121.16 m interval: numerous *Belemnitella*, predominantly extensively bored and corroded, from horizons within Glenarm α and Ballintoy B.

9 Owen (1970, p. 67) commented that typical examples of *Kingena pentangulata* (Woodward) had been collected from the Port More Borehole between the depths 411 and 500 ft (125.2 and 152.4 m), but that all the specimens appeared to be proportionately smaller than examples from English localities. It is now possible to give stratigraphical details of the records on which this statement was based, as follows:

422 ft 6 in (127.78 m): Ballintoy A1
433–436 ft (131.98–132.89 m): basal Ballintoy A1
448–450 ft (136.55–137.16 m): Larry Bane B
456 ft 5 in–458 ft (139.12–139.60 m): Larry Bane A
460 ft (140.21 m): Larry Bane A (low)
466 ft 6 in (142.19 m): Boheeshane C2
483 ft (147.22 m): Boheeshane C1
490 ft (149.35 m): Boheeshane B2

Of these records, all but the two lowest fall within the *Belemnitella mucronata* Zone s.l. The records in Boheeshane C1 and B2 are of interest in showing the downward extension of the range of this species into the higher part of the *Gonioteuthis quadrata* Zone s.l. (i.e. into the Lower Campanian).

10 Many thin-tested echinoids in the 140.76–141.07 m interval correspond to *Galeola*-rich horizons at the top of Boheeshane C2, immediately below the Boheeshane–Larry Bane boundary.

11 *Chlamys mantelliana* (d'Orbigny) was recorded from 141.43 m and 142.65 m, both depths falling within Boheeshane C2 (basal *mucronata* Zone). This species characterises the lower half of the *mucronata* Zone in Northern Ireland, England and north-western Germany (i.e. pre-North Antrim Hardground *mucronata* Zone and correlative successions) and is a particularly important guide fossil.

12 *Echinocorys* ex gr. *conica* was recorded from 145.08 m (i.e. Boheeshane C1), this being the lowest occurrence in the borehole.

13 The *Orbirhynchia* Band in Boheeshane B1 was proved between 153.06 and 154.76 m.

14 *Kingena blackmorei* Owen [NIE 7288: questionable determination] was recorded at an adjusted depth of 157.75 m[1] in Boheeshane A2. The significance of this record is discussed on p. 94.

15 *Coelosmilia regularis* Tomes was recorded at adjusted depths[1] of 157.88 m and 158.19 m in Boheeshane A2; at the latter depth it was associated with *Oxytoma* (*Hypoxytoma*) *tenuicostatum* (Roemer). This clearly marks the approximate position of the *Belemnitella praecursor* horizon (see p. 93), although no examples of the belemnite were recorded.

16 *Kingena lima* (Defrance) was recorded at a number of horizons within the Lower Campanian *Offaster pilula* Zone (Creggan Chalk), in addition to its usual occurrences within the Santonian crinoid zones.

17 The Basal Conglomerate at 168.35 m exhibits a thick laminated and glauconitised stromatolitic crust. Small irregular pyrite masses are also present.

STRATIGRAPHICAL RANGE CHARTS

Extensive collecting has been carried out as far as possible for each member from all accessible exposures within the sheet, supplemented by collecting from other relevant areas, notably the air-weathered cliffs of Rathlin Island in the adjoining Ballycastle (8) Sheet. Due to the excessive hardness of the sediment coupled with the effect of pressure-solution fossils are extremely difficult to extract, and the general standard of preservation is poor, especially where intergranular suturing has occurred at shell–matrix contacts. The collections are, therefore, somewhat unrepresentative, particularly in the case of those units where, due to inaccessibility or poor exposure, only limited surfaces were available for study. The most conspicuous fossils are echinoids, belemnites and brachiopods, although smaller elements such as bryozoa, serpulids, small corals, microbrachiopods and ophiuroid, crinoid and asteroid debris can be collected from suitably situated air-weathered surfaces. The brachiopods are almost invariably extracted in a more or less decorticated condition, and are consequently difficult to identify. With the exception of the three basal members, bivalves appear to be virtually unrepresented, but this is purely a function of collecting since core from the Port More Borehole yielded many more bivalves than were collected from the corresponding units at outcrop.

The palaeontological data are presented in the form of two range charts (Figures 23 and 24). The first covers the pre-North Antrim Hardground *mucronata* Zone plus the upper part of the restricted *quadrata* Zone; the second relates to the post-North Antrim Hardground succession up to and including the poorly accessible Maastrichtian beds near Portrush. Data from the Campanian–Maastrichtian boundary succession of Ballycastle in Sheet 8 (Wood, 1967) has not been incorporated as the exact correlation between the Portrush and Ballycastle successions remains to be established. In the range charts only significant and/or stratigraphically restricted species have been included, in order to emphasise the main faunal changes. In certain cases it has been necessary to adopt open nomenclature, pending revision of the groups concerned. Supplementary data for the lower part of the succession, and also for the units covered in the two range charts, is given in the form of faunal lists in the main text. It must be appreciated that the recorded vertical ranges are subject to the collecting limitations outlined above, and that subsequent work may necessitate their modification: the collecting from the post-Tanderagee succession is particularly unsatisfactory in this respect.

Fossil determinations were made by C. J. Wood, with the exception of the Bryozoa and Actinozoa, which were undertaken by A. W. Medd and A. A. Morter respectively. Mr N. B. Peake (Norwich) kindly examined a number of *Echinocorys*, and expressed an opinion on their taxonomic and stratigraphical position. The collecting was largely carried out by T. P. Fletcher and to a lesser extent by C. J. Wood.   T P F C J W

---

[1] A discrepancy exists between the original log and the detailed log of the lowest 11 m of the core prepared by Fletcher with respect to the Boheeshane A succession. The adjusted measurements referred to here have been derived by extrapolation from Fletcher's observations.

Figure 23
Vertical distribution of selected species in the upper part of the pre-North Antrim Hardground succession

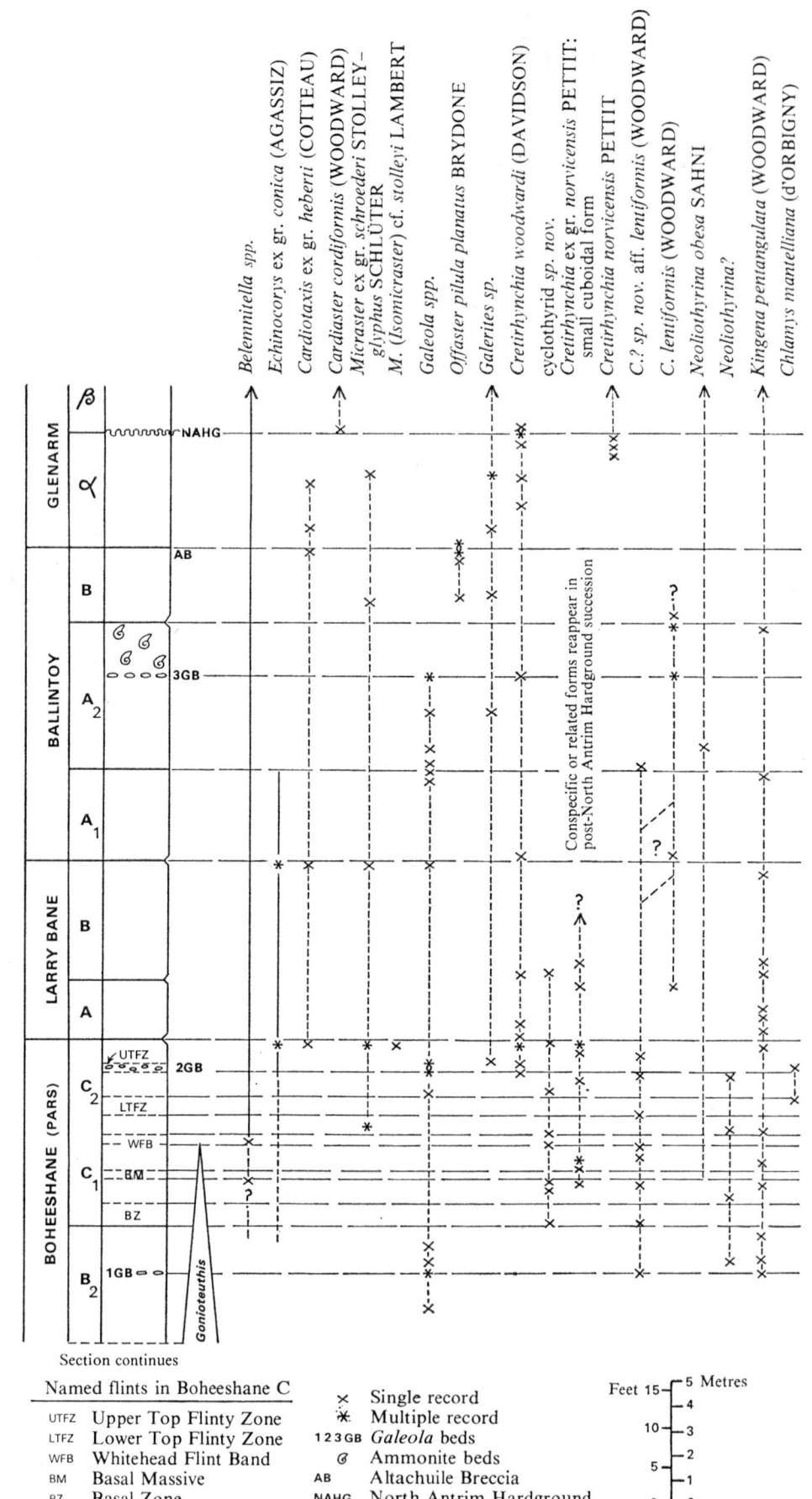

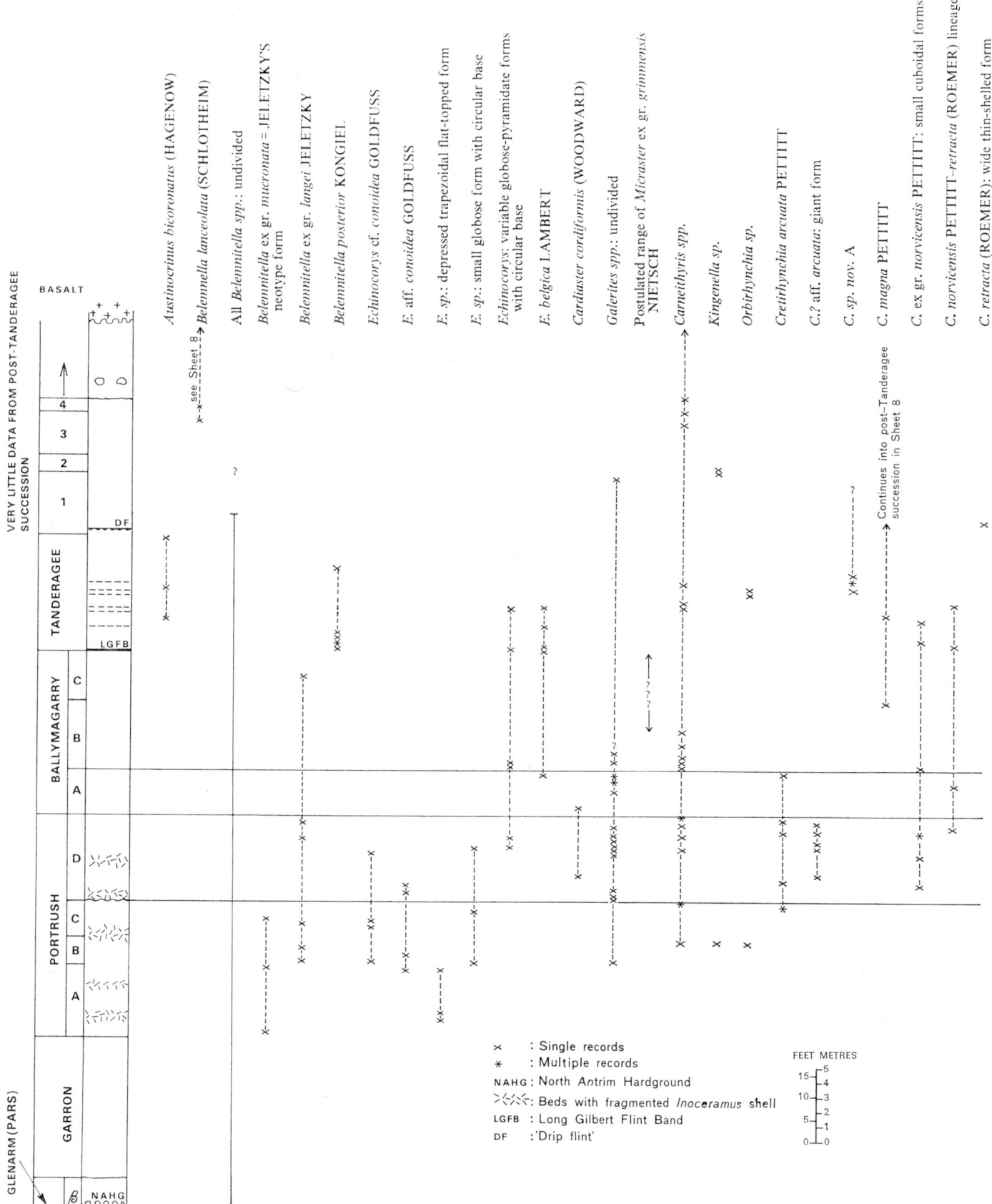

Figure 24 Vertical distribution of selected species in the post-North Antrim Hardground succession

CHAPTER 16
Tertiary extrusive igneous rocks

DETAILS OF EXPOSURES

Lower Basalts

The base of the lowest flow can be seen resting on the weathered Chalk surface in Ballymagarry Quarry [C 890 408] at the White Rocks 3 km east of Portrush. The 9 m of basalt seen is extremely rubbly and shows blocks of compact rock in a vesicular matrix, suggesting autobrecciation, or subsequent disturbance by the volcanic activity which has shattered the Chalk in the vicinity.

The Chalk/basalt contact in this quarry is irregular, with small swallow-holes and just east of the kilns it drops steeply into a conical depression 36 m across at the top and over 9 m deep, the base being covered by scree and fill. The depression is filled with rubbly basalt, not very different in appearance from that which overlies the rest of the quarry face. It has been suggested (Rohleder, 1926; Patterson, 1956) that this is a volcanic vent, but the occurrence of flint rubble below the basalt rubble, at the side of the roadway from the kilns to the east end of the quarry, indicates that it is more probably a large collapse or infilling of a hollow in the chalk surface. The latter supposition is borne out by the occurrence of basalt rubble below the main road in the seaward face of the quarry just west of Long Gilbert Bridge which suggests that the infilling may be of a hollow running north-north-west.

Where the base of the lavas is next exposed, in the cliff above Portnool 550 m to the east [C 896 410], the lowest flow, 6 m thick, is seen to be a compound lava with bands of pale grey-weathering rock alternating with bands of dark compact basalt. The grey basalts contain dark fresh basalt spheroids while the compact bands have irregular grey-weathering streaks. This flow is succeeded in the cliff by a massive flow of rudely columnar basalt 10 to 12 m thick with a thin reddened top, above which is a 6 m flow of rudely columnar lava with a central band of purple-weathering material. At Berginan's Port, 180 m farther east, this is in turn overlain by a 12-m flow of rubbly-weathering basalt.

The extremely impersistent nature of this series is shown by the section in the cliff behind Gulls' Point [C 900 411] where the well-marked succession at Berginan's Port, only 180 m away, is replaced by 15 to 21 m of rubbly basalt, sometimes with a banded structure, overlain by 12 m of massive rudely columnar basalt. At The Burnfoot the cliff above the Chalk displays a basal lava 15 m thick, massive and rudely columnar in its upper part but broken and decomposed in the 4.5 m above the contact. This is overlain by a second massive, rudely columnar flow 18 m thick and this in turn by 6 m+ of horizontally banded weathered basalt with compact and grey vesicular layers. This succession is cut off by a fault with downthrow to the east, and the cliffs and foreshore from Bonagarry to Dunluce are in five or more flows of extremely rubbly and decomposed basalts irregular in thickness and degree of weathering. The sea-washed foreshore exposures are of masses of coarse-grained vesicular basalt embedded in a greenish matrix, occasionally with laterite flecks. Though the Chalk is seen below high-water mark in Portnalea, presumably bounded by faults so small that they make no appreciable shift of the lavas in the cliff behind, there is no sign in this succession of the two massive lavas seen at Burnfoot, only a few score metres away.

The promontory on which Dunluce Castle sits has been described (Patterson, 1956) as a basaltic vent agglomerate. Though the existence of such a vent is accepted in the area of 'The Pound', on the neck of the promontory, the Castle rock itself is essentially similar in appearance to the cliffs behind Portnalea, and the Chalk on the foreshore to the east appears to pass beneath the base of the lava pile. It is therefore suggested that the Castle promontory is an outlying mass of the lava succession and not vent agglomerate. Immediately overlying the Chalk, however, there is an irregular layer of spheroidal basalt up to 6 m thick which is probably an agglomerate though it might be a ball lava. Patterson (1962) obtained blocks of altered chalk from this layer. It is bisected by a north–south fault of no great throw along which a cave has been cut right through the rock. The survival of this projecting spur of rubbly and easily weathered basalt is due to the occurrence, at sea level and just below it, of the Chalk which here is more resistant to marine erosion than the lower beds of the lava succession.

The bay east of the Castle is traversed by a north–south fault probably with appreciable downthrow to the east. It cuts off the Chalk on the foreshore north of The Sugarloaf, and brings in lavas which are presumably higher in the series than those seen to the west. The most conspicuous feature here is a composite (pahoehoe) flow of variable thickness (24 m+ at the Sugarloaf: 13.5 m at Gortnabane) with layers of dark compact coarse-grained basalt and pale grey kaolinised vesicular basalt, often with calcite veins and residual spheroids of less decomposed basalt and sometimes with thin undulose sheets of fairly fresh, fine-grained basalt up to 15 cm thick. It is seen in the cliffs and foreshore from Dunluce to Portballintrae, and closely resembles the composite flow seen on the north side of Rathlin Island (Wilson and Robbie, 1966, p. 192).

The composite flow is overlain and underlain by fairly massive flows which are seen at Gortnabane where the section totals over 76 m.

Cliff section at Gortnabane	m
Basalt, fine-grained, rubbly-weathering; seen above the road	9.0+
Lithomarge, purple and red with relict basalt spheroids; seen above the road	1.8
Basalt, massive, rudely columnar	~6.0
Basalt, composite flow, dark and pale-weathering bands: thin reddened top	9.0
Basalt, massive, rudely columnar, reddened top. Base incorporates a slab of laterite	4.5
Basalt, rubbly, vesicular, with lateritised and lithomarged top	9.0
Basalt, rudely columnar, coarse-grained. Lithomarged top 3 m thick	10.7
Basalt, composite banded flow	13.5
Basalt, massive, coarse-grained, irregular top, reddened in places	9 to 12
Basalt, massive, coarse-grained, 0.9 m laterite band at top	6.0+

The massive basalt below the composite flow is seen in the rock platform at Sheep's Isle, Keevenagh and White Shoulder [C 915 421] while the composite flow is generally in the cliffs. The overlying lavas are seen in the cliffs as far as Portnacapple [C 912 418] but to the east of this point they are only occasionally exposed. South of Keevenagh, however, the 7.5 m of lithomarge and basalt spheroids which mark the top of the banded flow are overlain by a 15-m flow of rudely columnar lava, deeply weathered in its upper half, and this in turn by two flows 4.5 m and 3 m+ thick respectively.

The raised-beach stack of Lacknamodeen consists of a mass of blocks of compact and vesicular basalt in a lateritic matrix. Rude bedding in the stack seems to dip south at 30°. Over an area 270 m across, centred on Lacknamodeen, the basalts are seen to dip concentrically towards the stack. It has been suggested (Patterson, 1962) that this is a volcanic vent but it is possible that the whole feature is due to collapse into a solution hollow in the underlying Chalk.

The Lower Basalts underlie the area between the coastal cliffs and the prominent scarp formed by the more massive Tholeiitic Basalts to the south, and form a series of impersistent and generally small scarps which appear through the drift cover in the townlands of Clooney, Dunluce, Boneyclassagh and Ballintrae. To the south-west, in the townlands of Ballymagarry, Ballymacra, Ballykeel, Ballycraig and Cloghorr, the drift cover is scanty in places and there are extensive exposures and many small scarps formed by the flows of this series. One of the more massive and compact flows near the top of the series has been quarried in a number of small excavations just south of the minor road in Ballykeel and Bally-macrea Upper, the greatest thickness seen in worked faces being 3.7 m. The flow above this, exposed in crags along a well-marked scarp, frequently displays good flow-banding, etched out on weathered surfaces. The lowest flows seem to be generally thin and impersistent. The cutting formerly occupied by the quarry tramway at Craigahullier shows about six thin slaggy flows, below a massive flow 9 m thick, seen at the tunnel under the minor road. This flow, with a 1-m laterite bed at top, is overlain by a 6-m+ flow to the south. The small stream which flows north-westwards from Upper Burnside [C 872 389] reveals about five thin flows in 180 m, and, 2 km to the north-east, but at about the same stratigraphical level, the road cutting at Ballymacrea Lower [C 885 404] shows the following section:

Road cutting at Ballymacrea Lower	m
Basalt, medium-grained	1.8+
Basalt, medium-grained, vesicular in upper 1.8 m with 15 cm to 0.9 m of red laterite on top	6.0
Basalt, massive, vesicular in upper 1.8 m with 15-cm purple lithomarge on top	6.0
Basalt, medium-grained, vesicular; pipe amygdales at base	1.8
Basalt, medium-grained, irregular thickness	~1.8
Possible fault	
Basalt, medium-grained, spheroidal weathering	1.1+

In the area south of Ballywillen there are a few extensive exposures of Lower Basalts between the outcrop of the Inter-basaltic Bed and the Islandmore Fault. These exposures of massive coarse-grained basalt are in the upper flows of the series. A borehole at the eastern side of Corbally reservoir [C 880 387] started just below the Interbasaltic Bed and penetrated 107 m of basalt 'frequently red-stained and amygdaloidal', with some slickensiding at 100.5 m which may indicate a fault at that point. It then cut 12.2 m of soft green pyritous material described as tuff-like, underlain by 23 cm of lignite. The next 33 m in the borehole was basalt, the driller's log having no comment on amygdales or reddening, and as the beds which cut the chalk through the next 107 m of the borehole were also described simply as 'basalt' it is probable that the band penetrated between 119 m and 152 m is in fact intrusive dolerite.

This borehole, then, suggests that the Lower Basalts of Corbally are almost 120 m thick and include a band of tuff 12 m thick at the base of the succession.

CAUSEWAY AREA

West of Leckilroy Cove [C 937 440] the low cliffs above the raised-beach rock-platform are capped by a massive rudely columnar lava over 12 m thick which overlies two thinner flows, 4.5 m and 6 m thick, each with a slaggy weathered top.

East of the Portcoon fault a number of fairly thick flows overlie a series of very thin and irregular lavas, with a general dip to the south-east.

Cliff section Portcoon–Portnaboe	m
Basalt, compact	3.0+
Basalt, irregular flow with slightly reddened top. Irregular bands of purple vesicular material through the whole thickness	10.7–13.7
Basalt, very massive, compact, medium-grained with red laterite bed 0.3 m thick at top	9.0–10.7
Basalt, coarse-grained, spheroidal weathering. Irregular top with patches of laterite and lenses of lithomarge for as much as 6 m from top	10.7
Basalt, massive, coarse-grained. Thin laterite bed at top	1.0–1.8
Basalt, vesicular, coarse-grained. Lithomarged and reddened top	1.8
Basalt, vesicular, coarse-grained. Lithomarged and reddened top	0.6–1.5
Basalt, vesicular, coarse-grained. Lithomarged and reddened top	1.8+

It is probable that the topmost flow of this succession is that seen again on the foreshore east of the boat slip at Portnaboe [C 942 442] where an eastward succession of thin flows passes under the succession displayed on the Great Stookan [C 943 446]. The complete section is as follows:

Lava succession in Great Stookan and Portnaboe	m
Basalt, rudely columnar, medium-grained, forming top of Great and Little Stookans	7.6+
Basalt, rudely columnar, coarse-grained, with slaggy top, sometimes lithomarged	6–9
Basalt, spheroidal weathering, coarse-grained, dying out at north	0–3
Basalt, coarse-grained, with a band of horizontal jointing about the middle. On the east side of Great Stookan the flow is confused by several converging and diverging purple vesicular lithomarge bands The base of the flow is very irregular and the top is lateritised	7–9
Basalt, massive, coarse-grained with thick slaggy lithomarged top. Dies out at north	0–4.5
Seen at Great Stookan:	
Basalt, very vesicular, lateritised. Top and base very irregular	1–3
Basalt, massive, coarse-grained. Irregular lithomarged top	6.7+
Seen in Portnaboe:	
Basalt, medium-grained. Bright red lateritised top	3.7
Basalt, coarse, spheroidal weathering. Deeply weathered and lateritised	3.7
Basalt, coarse, spheroidal weathering. Deeply weathered and lateritised	1.8
Basalt, massive, coarse-grained. Lithomarged top	~9.0

The equivalence of these lower beds at Great Stookan and Portnaboe is uncertain because of the known impersistence of the thin flows. Where the second highest flow is seen at Windy Gap beside the path to the Giant's Causeway it exhibits spectacular spheroidal weathering (Plate 14).

The flows seen on the foreshore at Port Ganny [C 946 445] underlying the purple lithomarge of the Interbasaltic Bed, are doubtless the extension of those seen at the top of the Great Stookan succession, but precise correlation is impossible. It is probable that the lowest lava, poorly exposed among the beach boulders at the west end of the bay is the fourth flow from the top of the Stookan series.

Lava succession in Port Ganny m
Interbasaltic bed
Lithomarge, reddish, passing down into coarse-grained basalt 1.8
Basalt, close-jointed, coarse-grained. Purple lithomarged top 1.8–4.5
Basalt, massive, coarse-grained. Lithomarged top ~4.5
Basalt, massive, coarse-grained. Lithomarged top ~9.0
Basalt, massive, coarse-grained, poorly exposed ~9.0

The correlations suggest that the Lower Basalt succession from Port Ganny to Leckilroy Cove is approximately 90 m thick and includes about sixteen separate flows with average thickness of 5.5 m.

From Port Noffer [C 950 447] to Benbane Head the upper part of the Lower Basalt series is well exposed in the cliffs below the Interbasaltic Bed, which rises from below sea level at the Giant's Causeway to 30 m OD at Roveran Valley Head and maintains that level at Port na Callian Head, rising, partially by faulting, to about 60 m OD at Benbane Head.

The cliff at Roveran Valley Head and the Amphitheatre shows the topmost lava to be a massive, hard, medium-grained basalt, sometimes showing flow-banding on weather-etched surfaces. Only 6 m or so thick at Roveran Valley Head, it becomes more massive to the east and is over 11 m thick above Spaniard Rock. Further to the east it can be followed as far as Plaiskin Head where it is about 7.6 m thick, though here it is separated from the Interbasaltic by a 6-m lava which comes in at Port na Callian.

At Port Reostan the persistent flow is underlain by a very vesicular and rubbly lava, with 1.5 m of lithomarge at the top, which thins from 7.6 m at Sea Gull Isle to disappear on the west side of the Amphitheatre. This in turn is underlain by two thin flows of compact, fairly coarse-grained basalt with lithomarged tops. The lavas, 4.6 m and 2.4–3.6 m thick respectively, are seen as far as Port na Spaniagh where they are 6 m and 2.4 m thick, the lower having become very slaggy. Beyond the western end of Port na Callian they are not recognisable and appear to be replaced by a slaggy flow 4.6 m thick, still seen at Plaiskin Head where it is reduced to about 3 m.

The massive lava seen at sea level at Sea Gull Isle, where it is over 12 m thick, is seen at Spaniard Rock to be 15 m thick, with a characteristic deeply weathered top, generally with a thin bed of laterite. It is seen at Lacada Point where it forms the promontory and the lithomarged top shows veins of laterite well below the upper surface, and can be followed to Plaiskin Head where it is still 10.7 m thick, and is presumably one of the upper flows seen at Benbane Head, having maintained its thickness for over 1.6 km. This lava, which may be called the Lacada flow because of its accessible development at that point, is underlain at the Amphitheatre by a massive flow over 6 m thick with a bright red laterite top. It is not certain whether the flow seen at sea level below the Lacada flow at the Giant's Eyeglass is the same lava but it also has a reddened top.

The best section in the Lower Basalt series is that at Plaiskin Head where the succession is:

Lower Basalt lava succession at Plaiskin Head m
Interbasaltic Bed
Basalt, massive, irregularly jointed 6.0
Basalt, massive, irregularly jointed, red laterite top 7.7
Lava, thin, slaggy, generally grassed over 3.0
Basalt, massive, irregularly jointed (Lacada flow) 10.7
Basalt, massive, irregularly jointed, with thick slaggy top, slightly reddened 13.7
[On the east of the Head this flow thins to ~4.6 m and is overlain by a thin slaggy lava ~4.6 m thick which dies out to the west.]
Basalt, massive, coarse-grained, seen at sea level on west side 15.0
Basalt lava, thin, with reddened top seen only at Port na Plaiskin 3.0
Lithomarged top of a lower flow seen below HWM

To the east the flows seen at the Murder Hole are generally similar to the Plaiskin Head succession, with a 6 m massive flow at the top, three rather slaggy flows totalling 18 m and two massive 15 m flows at the bottom of the cliff.

At the headland of Benbane only five flows are seen above sea level in a thickness of 67 m, one of the uppermost flows having disappeared and the lowest two having thickened to 21 m each. In the eastern side of the Head, however, two thinner flows are seen and the upper of the massive lower flows has thinned out to 9 m. The succession is:

Cliff section in Lower Basalts west of Port na Truin m
Interbasaltic Bed
Basalt, massive, irregularly jointed, medium-grained 7.6
Three massive but irregular flows with slaggy tops.
  One dies out to west 30.0
Basalt, massive, coarse-grained, with laterite top 9.0
Thin slaggy flow 4.5
Basalt, massive, coarse-grained, irregularly jointed 9.0+

It is notable that the promontory of Benbane Head, the northernmost point on the Antrim mainland, is formed by the local occurrence of two unusually massive and resistant lavas. Where these die out or are brought by the general southerly dip of the series (about 4°) below sea level, the marine attack has been able to advance further south.

East of Port na Truin the southerly dip brings the top of the Lower Basalts to sea level at Port Moon. Just south of Contham Head four thin flows averaging about 3 m overlie a massive lava 11 m thick, forming the raised-beach rock platform.

To the south of the Portbraddan fault and away from the coast the Lower Basalts underlie a wide area but exposures are few and usually poor. Massive coarse basalt is seen in a quarry near Craig Cottage [C 998 435] and there are scattered crag exposures in the Ballynastraid area. East of this the outcrop narrows rapidly and south of White Park Bay the base of the Interbasaltic Bed is only a few metres above the top of the Chalk, indicating that at this locality there is probably only one lava present in the succession.

In the area around Ballintoy the succession increases in thickness, and the lavas are underlain by chalky tuff, derived from the vent at Carrickarade or perhaps, as the volume of chalk appears to be so enormous, from an associated vent offshore. The western limit of the tuff appears to be about Mount Druid, but there are few exposures in this area. In the area south of the Church the scarp above the road is about 80 m high, of which about 40 m appears to be tuff and 30 m lavas, with the Interbasaltic Bed coming in at 150 m OD.

As seen in the stream section at the Fair Green, at the east end of Ballintoy village, the tuff rests on the top of the Chalk and the lowest 0.6 m is dark brown and heavily charged with manganese. Tomkeieff and Patterson (1948, p. 207) ascribe this manganiferous bed, which may reach 2 m or more thick, to penecontemporaneous bog formation but it seems more likely to be due to later deposition by groundwater.

The bulk of the tuff bed is pale in colour and consists predominantly of chalk fragments, usually small but locally large enough to make the rock an agglomerate, as on the west side of Knocksoghey Hill. The matrix of comminuted chalk contains, as well as chalk debris, fragments of basalt and Lias shale and rare sandstone, which may be of Triassic age. The upper part of the bed is often deeply lateritised.

The tuff reaches its maximum thickness of about 55 m in the area just east of Ballintoy and thins eastwards to about 25 m at Portaneevey on the east side of Carrickarade where it is underlain by a single lava 9 m thick. The upper flows of the series here total 70 m thick, so the lava succession thickens rapidly to the east of Ballintoy village.

In the cliffs from Portaneevey to the Giant's Cut the tuff bed, still with an underlying lava, thins gradually. At the Giant's Cut (Plate 25) three lavas, one a pahoehoe-type banded flow, totalling some 18 m overlie a tuff bed 7.5–9 m thick, of which the uppermost 2 m or so are lateritised. A single flow of blocky-jointed basalt 8–9 m thick underlies the tuff and rests on the top of the White Limestone.

North of the Ballintoy Fault a small outlier of the lava series is seen in stacks on the foreshore and raised-beach platform west of Ballintoy Harbour (Figure 7). The basal tuff bed, which here consists of some 9 m of chalky tuff, lateritised in the top metre or so, resting on over 5 m of coarse agglomerate with basalt boulders, is seen on the foreshore south-west of the harbour. Near Stackandoon [D 035 452] these beds dip north-west at up to 40° away from the fault.

The tuffs are overlain by basalt lavas which underlie the quay and form three small stacks in Portnalug. The lower of the two flows seen thins out on the foreshore, the underlying lateritised tuff apparently merging with the lateritised top of the flow at low-water mark beside the eastern stack. Beyond this there appears to be only a single flow of Lower Basalt about 5 m thick below the Interbasaltic Bed, but the area is below water and there may be faulting which would alter this conclusion.

Where seen at the western end of Port More the tuff bed is less than 10 m thick. The Port More fault, with downthrow to the east, brings the tuff to near sea level and a gentle fold takes it below low-water mark for some distance. At the eastern end of Port More, however, it reappears 9 m thick and is continuous with the complex of tuffs, locally underlain by a single lava, at Kinbane (see Ballycastle Memoir). The cliffs here show the Lower Basalt succession, including the tuff bed, to be some 107 m thick and the tuff rests directly on the White Limestone.

In the Port More Borehole, which was drilled at Glenstaghey about 700 m south of the Giant's Cut, the tuff bed was 6 m thick and rested directly on the Clay-with-Flints above the Chalk. The basalts were 70 m thick and, as the borehole was started some 10 m below the top of the Lower Basalt succession, the thickness of the group here is about 87 m.

Interbasaltic Bed

The westernmost exposure of this bed is in the abandoned mine adit in Crossreagh townland [C 873 381], 550 m north of Islandmore Upper village. The position of the outcrop here is fixed by the conspicuous scarp of the Tholeiitic Basalts as far south as the disused quarry, to the south of which it is displaced by a small fault and runs south-west till cut out by the Portrush Fault. The mine section is now almost inaccessible and only about 0.6 m of red ferruginous laterite was seen below the tholeiitic basalt in 1957. The ore was known but not worked at the time of the 1886 survey, but by 1912 Cole (p. 20) described the adit as 'old' and makes no mention of the section seen. A field map of W. B. Wright dated 1916 describes the ore as 'undoubtedly good and 2–3 ft thick at least'. In 1941, V. A. Eyles noted that below the basalt roof a few centimetres of light clay overlies up to 0.9 m of red pisolitic iron ore, moderately hard but very crumbly. The top of the bauxite only was seen, and is described as 'very poor—more like mixed pavement and lithomarge'. 'Islandmore', presumably this mine, is included in the list of working mines in 1902–3 and 1905–6, producing iron ore.

Red 'ore' and grey and reddish 'ashy beds' were met in excavations at Ballywillan Reservoir, 400 m north of the mine. Beyond this the outcrop was shifted almost 800 m eastwards by a fault and runs along the east side of Corbally reservoir and northwards to Craigahullier Quarry [C 880 390] (Figure 25). Workings in the tholeiitic basalt at this point have revealed a hollow in the Lower Basalts about 24 m deep and 135 m long, deepening from north to south. In this hollow there was an accumulation of lignite which thickened towards the bottom, finally reaching a thickness of about 6 m and is said to have stopped against a vertical face of large boulders and clay, probably relict spheroids in lithomarge, at the point now occupied by a disused crusher house. The deposite of lignite was almost completely worked out and only a small patch on the west side of the hollow was seen at the time of the survey (1957). About 1 m of hard black bright conchoidal-fracturing lignite overlain by columnar tholeiitic basalt rested on purple-grey lithomarge with spheroids of basalt and impersistent lenses of red laterite at the top. The whole dipped steeply eastward at 40°.

The lignite was described by Patterson (1946, pp. 437–8), who gave a chemical analysis of the material exposed when he visited the locality, as dull, brownish black, rather shaly, with a poor conchoidal fracture:

Chemical analyses of Craigahullier lignite and residual ash

Lignite				Ash	
	1	2	3		4
Moisture	9.34	8.55	8.6	SiO_2	39.02
Volatile matter	29.54	44.80	31.9	Al_2O_3	41.28
Fixed carbon	17.29	16.80	15.3	Fe_2O_3	8.84
Ash	43.83	38.44	44.2	FeO	–
				MgO	1.48
				CaO	4.69
				TiO_2	0.50

Analyses by: 1 Patterson, 2 Belfast Gas Works, 3 Coal Survey Laboratory, 4 Patterson.

Patterson suggested that the hollow infilled by the lignite was formed by a river, but this seems impossible as we now know that it deepens steeply to the south as well as from east and west, but ends abruptly after 135 m or so, the outcrop of the Interbasaltic Bed apparently continuing at an even level round the face of the quarry south of the crusher. The cause of formation of a hollow of this type is obscure, but it might have been a pit crater of the type seen in Recent lavas in Hawaii.

Bauxite and possibly iron ore were worked from 3 adits in this area prior to the publication of the 1912 Memoir though no indication of workings is given on the field map of 1886. Craigahullier was mentioned as a bauxite producer in the returns of 1906. When visited by Cole (1912, p. 20) the adits were 'long disused' and lignite, red clay, 'pavement' and lithomarge were seen on the old tip. The northern adit showed 0.6 m of lignite resting on 0.3 m of earthy hematite. In 1913 R. Welch described the works at the quarry in a letter to the Director of the Geological Survey and at that time 1.35 m of black coal-like lignite was exposed at the west side of the quarry face. A report of a Mines Inspector (GJW) in 1916 shows that a mine at Craigahullier was

Figure 25 Locations of old mines in the Interbasaltic beds in the area south of Portrush

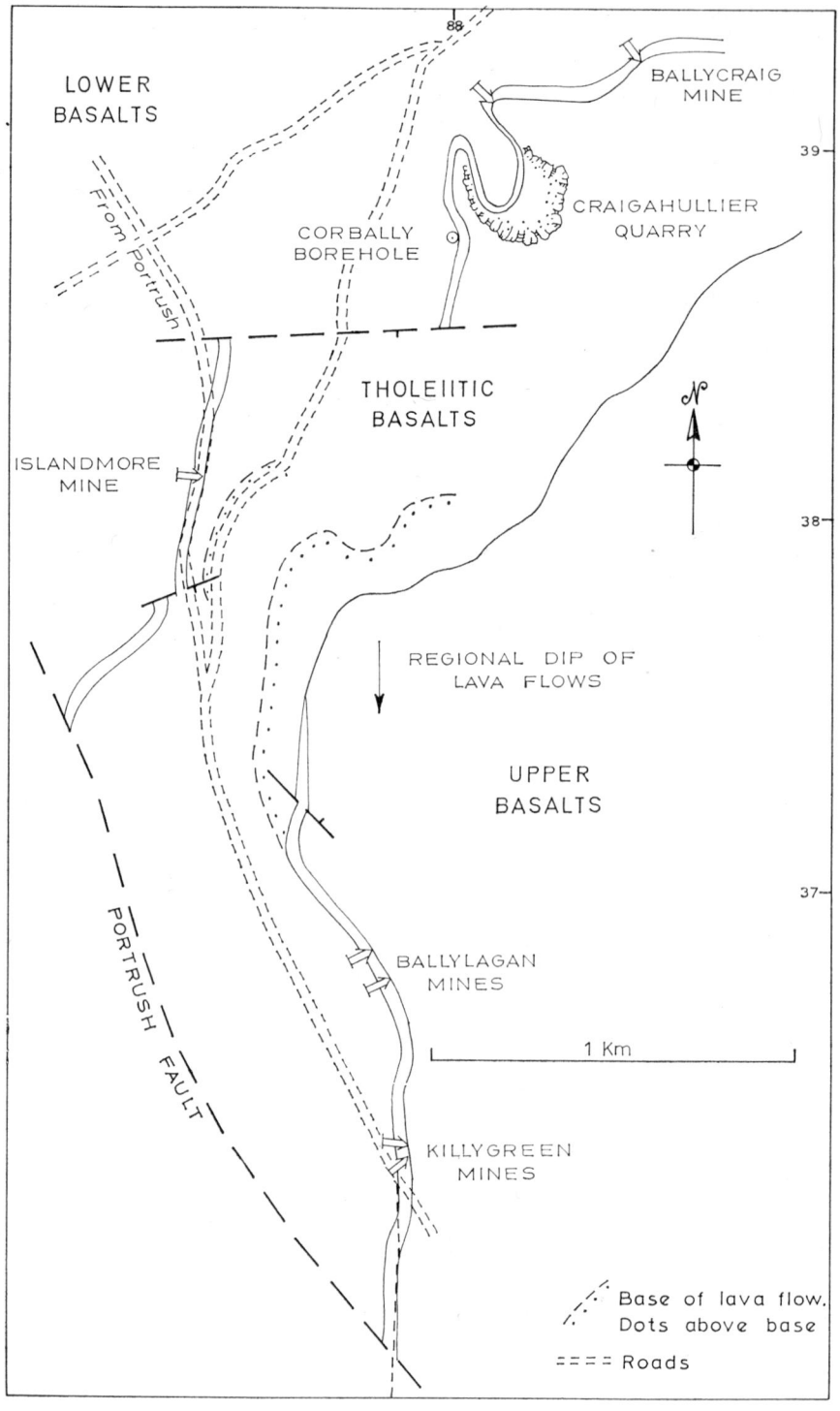

working at this time but whether it was producing lignite or bauxite is not known.

When the area was examined by V. A. Eyles in 1941 the small tip from the southernmost adit, only 90 m north of Corbally Reservoir, included grey bauxite like that from Ballintoy. The next mine, 135 m further north, was said to have worked bauxite or iron ore and grey bauxite debris was seen here too. The third adit, situated behind the offices and presumably Cole's 'northern' adit, was open, as it still is, and exposed bauxite at the mine mouth.

The section now seen in the east side of the quarry near this adit is:

Carbonaceous dirt	0.15 m
Bauxite, more or less ferruginous. Sometimes red but generally buff-coloured. No iron ore concentration at top	0.6–1.8 m
Lithomarge, purple and grey	up to 3.0 m

A chemical analysis of the bauxite from this exposure is given in Appendix 4.

East of Craigahullier the Interbasaltic Bed rises steadily to the disused mines at Ballycraig Upper [C 885 392]. Here two adits at different levels show an incomplete section through the Interbasaltic. The upper level (eastern) shows 0.07–0.3 m of dark-brown slightly pisolitic iron ore under the roof of tholeiitic basalt, passing down into over 2.4 m of bright red ferruginous bauxite with lithomarge nodules in the lower part. The lower adit, with top 1.5 m below the floor of the upper level, shows 1.5 m of red bauxitic lithomarge passing down into over 1.5 m of purple lithomarge. Subsidence over old workings is seen 90 m south-west of the mine entrances. These adits were described in the 1888 Memoir (p. 24) as iron ore mines, the only ones then working in the district. They are not mentioned in the 1912 Memoir and were disused when seen by Wright in 1916.

The Interbasaltic Bed is not seen for about a kilometre to the east-north-east, though a small excavation in Ballykeel townland may have been a trial pit or level. At Crocknamolt [C 893 396], Urbalreagh townland, a group of four old adits on both sides of the minor road are now almost completely obscured by falls and tippings. The 1886 field map shows only one level so the mines were worked after that date, but by 1916, when Wright visited the area, the mines appear to have been abandoned. V. A. Eyles noted in 1941 that a local resident said the mines had been worked for bauxite in 1912 and that the seam was about 0.9 m thick. Among the tip debris Eyles noted lignite and grey bauxite like that from Ballintoy. In 1957 only the westernmost adit was partly open and showed 1.3 m of red ferruginous bauxite under Boulder Clay.

About 640 m south-west of these workings two old levels are seen to the west of the main road. They are marked on the 1886 field map but no information about them is given. In 1912 (Cole, p. 21) they are described as long disused. Eyles, in 1941, noted 0.23 m of poor iron ore, some lignite and dark-red mottled bauxite. In 1957 0.15 m of slightly pisolitic ore resting on over 0.9 m of red ferruginous bauxite was seen below the columnar theoleiitic basalt which forms the roof of the northern adit.

The course of the Interbasaltic horizon can be followed over the next two kilometres only by the rather vague features formed by the tholeiitic basalts; and nowhere is there any exposure of the laterite. The next group of workings are at Glentask and an open adit near the farm [C 909 403] shows the following section:

Section at Glentask Mine	m
Basalt, irregularly jointed, tholeiitic	1.5
Basalt, decomposed	0.3
Iron ore, red, slightly pisolitic	0.3
Bauxite, red, ferruginous, with increasing streaks and nodules of lateritic lithomarge towards base	3.6

One of the adits is marked on the 1886 field map and in the 1888 Memoir it is stated that 'lately, however, rich deposits [of iron ore] have been opened south of Dunluce Castle'. 'Dunluce and Glentask' is included in the list of producing mines in 1905–6 but Cole (1912, p. 21) states that in 1907–8 they were closed but were said to have been recently opened again. Wright's field map of 1916 suggests that the mines were not then working but were in good condition. A second adit at Glentask is now obscured though Eyles, in 1940, was able to see about 0.3 m of iron ore and 0.6–0.9 m of ferruginous bauxite. A third adit 270 m north-north-east of the Glentask pair and just inside the townland of Dunluce is now completely fallen in and no section is seen. Wright, in 1916, noted 0.46 m of slightly pisolitic iron ore or 'pavement' at this point.

Beyond the Dunluce mine the outcrop runs round the hill of Mullaghintorp at about 76 m OD being affected by a small fault south of Bonnyclassagh farm and a larger fault west of Ailsacraig. Old trial pits are still seen at Nancy's Lope [C 911 413] and south of Bonnyclassagh. At the former, the lowest flow of the tholeiitic basalts is only 3.66 m thick and overlies pisolitic iron ore only 0.02–0.05 m thick over 1.2 m+ of ferruginous bauxite. Just east of the latter, dark-red ferruginous bauxite is seen in situ at two localities with traces of lithomarge further down the bank.

From Ailsacraig the outcrop of the Interbasaltic Bed has a sinuous course beneath the town of Bushmills and across country, stepped by the series of north-north-westerly faults in the Bush valley and marked only by the broken feature formed by the lowest flow of tholeiitic basalt. Exposures are few and inconclusive for about 8 km—considerably more along the outcrop—being confined to a few small shows of lithomarge and laterite south of Lisnagunogue and Ballyallaght.

From Ballynastraid to the eastern edge of the sheet the position of the Bed and its nature are much more clearly known as the result of workings during the last hundred years. In this area the usual ferruginous bauxite and iron ore found at this horizon are replaced by siliceous bauxite with a good deal of lignite. This indicates that the material weathered here was not the topmost lavas of the Lower Basalts but probably a rhyolitic or mixed tuff; of which no unweathered material remains.

The following notes are compiled from the original six-inch sheets of 1883–5, a later revision by Cole and others, the work of V. A. Eyles of the Geological Survey during World War II in 1940–43, some of which has been published (Eyles, 1952) and the recent revision of Sheet 7.

All trial pits, adits, and shafts have been serially numbered 1–36 from west to east, and are shown on Figure 26.

The bauxite horizon appears to be restricted to a belt extending from Lemnagh More to Ballintoy Quarry and though many pits are recorded, by no means all were sunk for bauxite, but either for iron ore or lignite.

The two mines exploited for bauxite are Lemnagh More and Clegnagh. No abandonment plans appear to have been made for either mine.

The Lemnagh Mines were formerly worked by the Eglinton Chemical Company and were closed in 1913 or 1914.

The Clegnagh Mines were last worked by Messrs Blackwell of Garston, Liverpool, and closed about 1920.

In this account the following areas are described:
1 Ballynastraid area 4 Clegnagh Mines
2 Lemnagh More Mines 5 Ballintoy Mines
3 Lemnagh Beg area 6 Remainder of outcrop

BALLYNASTRAID AREA (Antrim 3 SE)

In the base of the tholeiitic basalt quarry on the boundary between Lemnagh More and Ballynastraid townlands, about 1 m of Interbasaltic Bed was dug into, only the top two or three centimetres of lignite being now visible. About 15 m to the north-west of the quarry a depression in a small field probably indicates a trial pit (1).

LEMNAGH MORE MINES

A group of adits and shafts lie around the spur of tholeiitic basalts east of Ballynastraid Farm (3–8). One shaft and four adits can still be located on the ground and the old hand-engraved Ordnance Survey map shows two more shafts of which nothing is known. A note on the 1907–8 field map describes the occurrence on a tip of 'red and brown bole, lithomarge, a black substance which is probably decomposed tachylite, and a grey material .... probably the refuse of bauxite .... No possibility of arriving at the thickness of these beds owing to falling in of the roof of the adit'. Eyles notes '10 in lignite on bauxite' and may have opened an adit to sample the bed (6). At (7) he refers to 'a well-constructed old mine 7 ft high, waterlogged to the mouth'.

Eyles also records a small pit with brown rather ferruginous bauxite overlain by a few centimetres of lignite at a point 50 m

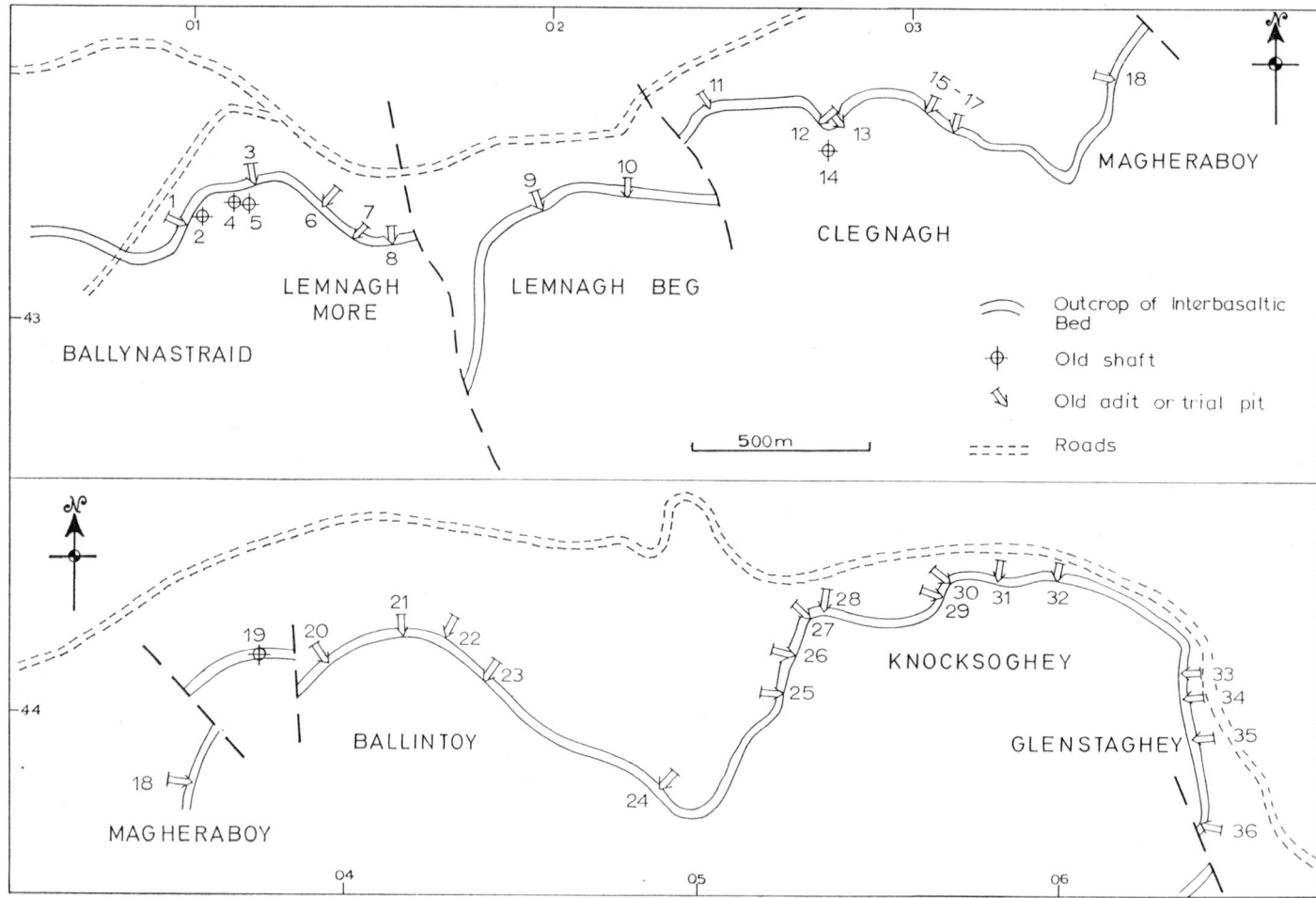

Figure 26 Locations of old shafts, adits and trial pits in the Ballintoy–White Park area

east of adit (6), old tip heap and made ground with many fragments of grey and pink bauxite between this and adit (6), and signs of bauxite between (6) and (7). He said of adit (8) that local information was that there had been a small mine there but that the bauxite quality was not good. There is no sign of a mine at this point and it is doubtful if there was any more than a small trial pit. Traces of ferruginous bauxite are seen at several points along the outcrop on both sides of adit (8).

Lemnagh Beg area

A series of trial pits are recorded along the outcrop across this townland but only one has clearly been opened as an adit. This (9) is still open but flooded and is used as a water supply to a local house. Eyles recorded fragments of grey and brown weathered bauxite but only a few metres of lithomarged basalt is now exposed. At 50 m east of this adit Eyles recorded an exposure of variable bauxite as follows:

'Grey Bauxite 2–3 ft, Purple rock (?lithomarge) 2–3 ft, Pink Bauxite 1 ft'.

About 100 m east of Lemnagh Beg farm Eyles noted an old mine (10) at which he saw 0.3–0.6 m of grey bauxite.

Clenagh Mines

Just north-west of Clenagh Cromlech a site on Cole's map showed 'at least 20 ft of red pavement and lithomarge under columnar basalt'. This may have been a trial adit (11) but nothing was seen by Eyles except traces of bauxite. He says, however, that this was used in working Clegnagh mine by Messrs Blackwell. No exposure is now seen. The main workings at Clegnagh were from two adits south-east of the farm, and an air shaft a hundred metres south (12–14). Entrance to the levels was still possible in 1940 and Eyles recorded the section seen as: 'Lignite and soft coaly matter 6 in to 1 ft; Hard red and purple rock up to 3 ft; Grey bauxite 2 to 3 ft'. Access is now blocked by falls but a dump of several hundred tons of unsold bauxite is still extant, though some has been removed for fill. The material is a grey bauxite with some free quartz crystals, and the analysis figures for this tip are given in Appendix 3.

Trial pits (15–17) to the east of the mine showed a thick bed of 'purple grey brown and red bole or lithomarge' but Eyles noted no sign of bauxite.

In the adjoining townland of Magheraboy lignite was seen in the bottom of a small quarry just east of the main road, and a small adit (18) 150 m west of the Cromlech was described as a mine for lignite. Another level 300 m east of Mount Druid is described on old maps as a 'coal adit' (19) but nothing is now seen here.

Ballintoy Mines

This group of adits and shafts (20–26) were all used to work the lignite, which was up to 0.6 m thick, about the beginning of the nineteenth century. So far as is known none of them ever worked bauxite or iron ore. The adits extend along the outcrop for a distance of about 1000 m. Westernmost is a 'coal adit' with a tip from which Eyles noted red bole and grey bauxite and where 1.5 m of red ferruginous lithomarge were seen during the resurvey. Almost 150 m north-east of this is an old adit, immediately west of a large quarry in the Tholeiitic Basalts. Eyles recorded an old shaft 300 m to the south, indicating that working here must have been on a considerable scale, but this site is not now located. Two further adits were sited on what is now the quarry floor, and another 75 m to the east a further adit 'to lignite. Spheroidal basalt on floor' is marked. Trial pits 'to ore' are recorded 150 m and 700 m further along the outcrop but no details are known.

It should be noted here that 'old workings' and 'trials' are mentioned on old maps at the top of the tuff bed and below the lower basalt lava about 165 m south of the main road in the area opposite the road from Ballintoy Port. These are at a lower level than the lignite workings and, so far as is known, never produced any ore.

At the eastern end of Ballintoy townland two old trial pits are no longer visible, but Eyles recorded 'dark red shaly fine-grained bole on altered lava' with little sign of bauxite.

Knocksoghey and Glenstaghey

A number of old trials are recorded along the outcrop in these townlands but there is little indication of what they found. A good section at the roadside at Ballaghacravey—just east of the entrance to the National Trust property at Carrickarade—shows the following section:

Section at roadside, Ballaghacravey	m
Columnar Basalt	
Coaly streak	0.10
Bole, dark red	8.23
Iron ore, dark red, pisolitic passing down into dark ferruginous kaolinised basalt with a trace of bauxite separating ore from lithomarge	0.30–0.45
Spheroidal basalt	

In Glenstaghey the outcrop swings round and runs north to south at the foot of a scarp feature. Five old trials (33–36) are marked on early maps of which the most northerly is still an open adit. Eyles records 0.1–0.15 m of poor-quality pisolitic ore over 1.1 m of red and red-brown lithomarge, overlain by tholeiitic basalt and underlain by spheroidal basalt. The thinning of the Interbasaltic Bed from 10 m or more at the Giant's Causeway to just over 1 m in this area is very striking.

The outcrop is moved north by the Port More Fault, with downthrow to the south-east, and can be followed along a feature across Coolmaghera. A small exposure of lithomarge at the extreme eastern edge of the sheet is all that is seen.

In the outlier at Ballintoy Harbour the outcrop is mostly below sea level. Indifferent exposures south of the islet of Dundonald show spheroidal masses of basalt in a brown zeolitic matrix with a concentration of iron pisolites at the top, apparently passing under the tholeiitic basalt. Red lateritic material seen about 20 m south of the contact is the weathered top of a lower flow and the Interbasaltic here seems to be represented by a partly kaolinised flow a few metres thick. The contact is not exposed, however, and may be faulted.

The bed is also said to be exposed on the foreshore near Dunnaglea but none of the observers who recorded this have described it and it is not exposed at present (1974).

In the outlier north of the Port Braddan Fault the Interbasaltic Bed is seen in cliff sections for several kilometres but only at the western end is it easily accessible.

The upper part of the Bed is seen beside the path from Causeway Head to the Giant's Causeway 550 m south of Great Stookan [C 945 441] where 4.6 m of red ferruginous laterite is seen below the tholeiitic basalt. It is also seen on the spur below Weir's Snout where the basalt rests on 1.8 m of dark red ferruginous laterite containing an increasing proportion of fragments or kernels of yellow kaolinised basalt towards the bottom and passing down into dark-weathering yellow bauxitic lithomarge with patches of dark ferruginous laterite, of which 2.4 m is seen. Intermittent exposures further down the slope for about 15 m show yellow bauxitic materials and lithomarge, most of which is probably in situ.

The level of the Interbasaltic Bed must now fall steeply to the east but outcrop is obscured by talus and the horizon is next exposed on the foreshore at The Highlandman's Bonnet [C 946 446] immediately west of the Little Causeway. Here there are extensive exposures of purple lithomarge with large residual spheroids of compact basalt, some of which stand on small undercut stacks of lithomarge to form little 'tam-o'-shanter'-capped pillars ('Highlandman's Bonnets'). The lithomarge is red- and orange-stained and vesicular for a metre or so below the overlying tholeiitic basalt of the Little Causeway, but the contact is not exposed and there is no sign of bauxitic material or iron ore. The low-tide promontory about 90 m west of the Little Causeway is caused by an increase in the number and size of these spheroids and marks the base of the uppermost identifiable flow of the Lower Basalts. The hollow to the west represents the weathered and kaolinised top of the next flow and exposures of laterite and lithomarge are seen among the beach shingle.

The Interbasaltic Bed dips beneath the Causeway and rises on the east giving a similar foreshore exposure at the west side of Port Noffer. Here the base of the Tholeiitic Basalts rests on 15 cm of dark-green fissile material which overlies about half a metre of red lateritic lithomarge, passing down into purple lithomarge. As in Port Ganny a concentration of residual basalt spheroids marks the base of the uppermost recognisable flow and the next lowest flow is also deeply weathered. In both these foreshore exposures only the lowest part of the Interbasaltic Bed is preserved, the upper and more friable ferruginous and bauxitic parts presumably having been removed by river action at the bottom of the valley in Interbasaltic times. The thickness of lithomarge must be about 23 m—thicker than is usual.

The bed rises to the east and, obscured by scree, is not seen again till it outcrops on the cliff path 140 m north of The Organ. A good section through the upper part of the horizon is seen on and above this path on the west side of Roveran Valley Head [C 952 450], where the weathered zone clearly includes two lava flows.

Interbasaltic Bed at Roveran Valley Head	m
Pisolitic iron ore	0.3
Lithomarge, purplish bauxitic	0.6–1.2
Lithomarge, orange and red lateritic with out-weathered irregular veins of hematite. Horizontal streaking with layers of purple lithomarge. Some partially kaolinised basalt spheroids, about 30 cm diameter, seen about 1.5 m from bottom, and completely decomposed spheroids seen from this level to the base. Grey lithomarge predominant in lowest 1 m	4.6
Laterite, red ferruginous—top of lower flow	0.15
Lithomarge, vesicular, purple, with vivid purple streaks and vesicles of white kaolinite	1.1
Discontinuous exposures of purple lithomarge in grassed-over slope	0.3

The lateritic beds are easily accessible in the cliff path for about 1 km beyond this point and beyond Benanouran Head are well exposed in the cliff, though often largely grassed-over. The thickness of the Interbasaltic decreases eastwards from some 15 m at The Amphitheatre to 9 m at Port na Spaniagh, and this thickness is maintained to Benbane Head with a local thickening at Hamilton's Seat to 15 m. At the cliff top in Port na Truin the Bed is about 9 m thick and is locally accessible. A small trial adit here shows about 15 cm of pisolitic ore at the top of the lateritic horizon. From here to Port Moon the Bed is seen in inaccessible section, often largely grassed-over, descending gradually towards sea level. There was another small adit at Portfad but nothing is now seen. Where the Interbasaltic reaches the sea at Port Moon it is completely obscured by scree and no exposures are seen.

Tholeiitic Basalts

The best sections in these lavas are in the coastal cliffs between the Giant's Causeway and White Park Bay.

West of Weir's Snout [C 945 443] the first flow of the series is about 18 m thick, with a slaggy top passing down into irregular starchy columns. The bottom of the flow is largely concealed by scree. The base of the second flow is seen near the cliff top east of Portnaboe, and consists of irregular starchy columns.

The cliffs behind Port Ganny [C 946 445] show the second flow to be over 15 m with a colonnade of irregular columns about 6 m thick passing upwards into thin starchy columns. The first flow is some 21 m thick at Weir's Snout where it is seen resting on the Interbasaltic Bed, but beyond this the base of the flow is concealed by scree. The visible part of the lava shows irregular columns above the scree level, only occasionally developing into a regular colonnade. To the west the upper part of the flow is massive and irregularly jointed but towards Aird Snout it becomes starchily columnar. At the eastern end of Port Ganny the purple lithomarge of the Interbasaltic Bed is seen on the foreshore and is apparently overlain by massive irregularly jointed basalt with rude columns plunging steeply to the west, though the actual contact is not exposed. The basalt, which is the base of the Giant's Causeway, is the bottom of the first flow of the Tholeiitic Basalt and it may be reasonably conjectured that the bottom of the flow and the underlying bed fall from about 61 m OD at the Causeway Hotel and Weir's Snout to sea level at the Little Causeway. On the east side of the Grand Causeway the base of the flow is again seen resting directly on lithomarge and it rises to the east beneath the scree slopes of Port Noffer to appear over 45 m above the sea at Roveran Valley Head.

In the cliffs east of Port Noffer [C 950 447] the first flow is 24–37 m thick. At the Giant's Organ the very regular colonnade is 12 m high and is overlain by 9 m of starchy columns and 12 m of massive irregularly jointed basalt, sometimes showing rude columnar structure. The colonnade is seen through a fan-shaped gap in the grassed scree and the massive columns, about 60 cm in diameter and dissected by cross-joints, do look like organ pipes. Here the base of the flow must be only a metre or so below the level of the path at the bottom of the columns. Where the base is again seen, at the headland 180 m north of The Organ, the colonnade is almost absent and the whole thickness of the flow, about 24 m, consists of starchy and irregularly jointed basalt.

About 180 m south of The Organ the slaggy top of the flow is seen, overlain by 1.5 m of orange-brown ferruginous laterite. Above this a thickness of about 1.5 m is grassed over before the base of the second flow is seen. It was from this position that lignite was worked and the bed is described as being 'upwards of six feet thick' (Portlock, 1843, p. 147). Nothing is now seen of this deposit.

The second flow of the series is well exposed in the cliffs from Aird Snout to Roveran Valley Head and reaches a thickness of 30 m above The Organ. It consists of a fairly regular 15 m colonnade with moderately well-developed columns, a thick entablature of massive irregularly jointed or rudely columnar basalt with only rarely any development of thin starchy columns.

The third flow in the succession is seen at the cliff top for about 275 m south of The Organ, where it consists of a thin colonnade and an entablature of thin irregular columns. The greatest thickness seen is about 9 m.

Above The Amphitheatre [C 952 452] the first flow has generally a columnar colonnade though it is absent at Hawk's Hollow. At the Irish Harp the columns are thin and curved in a fan-shaped aggregate. The thick entablature, 12 m or more, is of irregular starchy columns and the top of the flow is lithomarged and lateritised for a few metres. The total thickness is 24 m. The second flow has a fairly regular colonnade, part of which forms the isolated pinnacles of the Chimney Tops, and a slaggy entablature surmounted by a band of massive irregular columns. It is 20–24 m thick. The third flow seen in the south-east of the semicircle of cliffs, shows a maximum thickness of only 6 m, the lowest 3 m of which is a regular colonnade.

Behind Port na Spaniagh [C 955 454] the first flow has a regular colonnade 12 m thick overlain by a slaggy vesicular layer which is up to 7.5 m thick, west of the Spanish Organ. The rest of the entablature is of thin starchy columns, and is from 7.5–15 m thick, sometimes becoming massive at the top. The thickness of the whole flow is constant at about 27 m.

The second flow here is only 15 m thick at the west end but thickens to 27 m at the east, where a 12-m colonnade is overlain by an entablature of 6-m starchy columns passing up into 9-m massive irregular columns.

The third flow appears at Benanouran Head where it reaches a thickness of 9 m, all in thin starchy columns. It forms the cliff-top as far as Port na Callian Head where it swings to the south and is not seen again in the cliffs.

Beyond Benanouran Head the lowest lava thickens somewhat to the east with a colonnade 9–15 m thick overlain by a band of thin irregular columns 15 m thick at Port na Callian and 9 m at Plaiskin Head, and a top layer of massive irregularly jointed basalt 6–9 m thick. This three-layer division is particularly well displayed at Plaiskin Head where the flow is up to 37 m thick.

The second flow, now at the cliff top, reaches a thickness of about 21 m with a massive well-developed colonnade 10.7 m thick and a rather slaggy entablature. East of Plaiskin Head this flow also disappears from the cliff and forms a scarp across Tonduff Mountain [C 965 455] to the south-east. A quarry 275 m south-east of the headland shows over 9 m of massive fairly regular columns in this flow.

Benbane Head has only one flow of the Tholeiitic Basalts to cap it, here about 24 m thick with a colonnade 0–9 m thick overlain by a thick entablature of starchy columns.

On the dip-slope south of the cliffs higher flows of the series form scarp features across the townlands of Aird and Lisserluss. Crags of massive non-columnar tholeiitic basalt of the fourth flow run eastward approximately parallel to the cliffs from a point 457 m north of Aird Orange Hall for about 1.2 km and are then stopped by the Tonduff Fault. The fifth flow forms a scarp just 180 m north of Lisserluss and small quarries show massive non-columnar tholeiitic basalt resting, at one point, on 1.1 m of red lateritic lithomarge [C 966 445]. This flow is sparsely porphyritic.

A small scarp at Tonduff village may be the sixth flow of the series.

East of the Tonduff Fault the third flow forms a well-marked feature 135 m south of Plaiskin Farm and a short distance to the south of this, a small quarry [C 968 450] beside the lane shows 4.5 m of rudely columnar basalt of the fourth flow which can be followed eastwards to the cliff-top at Port Moon. The fifth flow forms a feature along the northern edge of Carrowreagh townland and eastwards to the cliff top above Portnahooagh and the sixth forms a scarp north of Carrowreagh Bridge but disappears eastwards under drift.

Along the coast south of Bengore Head, the southerly dip brings the lowest flow to sea level at Port Moon, and flows two to five can be traced in the cliffs between Benandir and Portnahooagh. The section here shows:

Flow 5   9 m
Flow 4 21 m
Flow 3 18 m
Flow 2 23 m
Flow 1 15 m+

The castle rock at Dunseverick is in Flow 5, but east of this an easterly dip brings this lava to sea level at Portnawhellan and the sixth, seventh and eighth flows successively onto the foreshore between here and Portninish.

Between Portninish and Port Callan a ninth flow of the Tholeiitic Basalts appears on the foreshore, below an outlier of Upper Basalts. Nowhere else in north Antrim are these uppermost flows of tholeiitic lavas known.

The inland outcrop extends from the Portrush Fault to the eastern edge of the sheet. In the west the lowest flow of the series forms a notable scarp feature east of the road from Ballywillan to Islandmore. A small quarry at Ballywillan Orange Hall [C 873 382] shows 3.7 m of columnar basalt, and in an abandoned quarry 270 m south of this very massive columns 7.5 m high and averaging 1.2 m in diameter pass upwards into a non-columnar slaggy top. The junction between columnar and non-columnar material is irregular and the latter, the top of which is not exposed, is at least 2 m thick. A flat terrace behind the quarry is backed by a well-marked feature which is probably a second flow, and suggests that here the first lava is not much over 9 m thick.

The underlying Interbasaltic Bed is cut out by the Portrush Fault some 550 m south-west of the large quarry and what is probably the first flow is exposed in two quarries which lie on opposite sides of the road 400 m south-west of Islandmore Upper. The northern face [C 871 371] shows 6 m of massive columnar lava while the southern, in the upper part of the flow, exposes 6 m of starchy irregular columns, with patches of non-columnar lava, overlying massive columns seen in the quarry floor. A number of other small exposures seen on the east side of the fault gully in the townland of Ballylagan are of tholeiitic basalt, but it is not possible to say to which flow they belong.

Scarp features in the relatively drift-free area of Crossreagh and Islandmore Upper suggest that there are at least two flows of tholeiitic basalt above the first. The lower of these is seen in small quarries 135 m east and 270 m south of Ballywillan Reservoir [C 875 383] which expose 3.7 to 4.6 m of irregularly jointed tholeiitic basalt, occasionally showing irregular columns. The other clear scarp which runs north–south 180 m east of Islandmore Upper and is now quarried immediately east of the village, shows massive rudely columnar basalt which, though tholeiitic, is noticeably coarser in grain than the usual rock of this type. Followed to the south, this scarp is seen to converge with that formed by the Upper Basalts just east of North Ballylagan, and rock of this distinctive type is not seen in outcrop south of this point. A residual spheroid of partially weathered rock taken from the Upper Interbasaltic Bed at South Ballylagan mine [C 878 366] is, however, of identical type and it appears that in this area the topmost flow of the series has been almost completely kaolinised and lateritised.

The extensive quarry at South Ballylagan Farm which exposes over 6 m of somewhat vesicular irregularly jointed tholeiitic basalt, and the occasional small exposure elsewhere between the Upper Basalt scarp and the Portrush Fault are thus presumably in the second flow of the series, unless there is an additional flow, or flows, which are not recognised from scarp features.

In the townland of Craigahullier an excellent section through the series is given by the workings in the Craigahullier quarry [C 881 389]. As described in the account of the Interbasaltic Bed, the lowest flow of the tholeiitic lavas here filled in a considerable hollow in the Lower Basalts and working in the quarry showed that columnar tholeiitic basalt continued down to the level now occupied by disused crushing plant, that is about 60 m OD, and there the first flow reached a local thickness of about 30 m. The workings now show this lowest flow to be about 9 m thick, and to be entirely irregularly jointed and non-columnar. The base and top are rather irregular and the uppermost 3 m is slaggy and vesicular and it is overlain by an irregular bed of lithomarge and laterite up to 1 m thick and sometimes containing streaks of lignite and pieces of carbonised wood. That this laterite bed is at least in part tuffaceous in origin is shown by the petrology of a specimen of red laterite (NI 2113), which is composed of glassy material and basalt fragments. The second flow is 9 to 12 m thick, with regular columns up to 3 m high in places but mainly with thin starchy columns or irregular joints. At one place, on the south side of the working, the base of the second flow falls into a small hollow, perhaps 18 m across, in the top of the first flow, and in this area a range of very regular small columns, up to 3 m long, has formed. The top of this flow is irregular and contains small drusy cavities. Weathering and purple lithomarge have penetrated up to a depth of 2.4 m in places and there are wisps and streaks of dark ferruginous laterite in places at the top of the bed.

The third flow is of massive fairly regular columns and, where seen in the working face, is over 9 m thick. This flow forms a very well-marked scarp feature along the hillside to the north-east of the quarry. About 275 m from the workings, a small area of subsidence or an old trial shaft is seen at the foot of the scarp and 90 m beyond this, in the townland of Ballycraig Upper, two adits [C 885 392] have been driven into the bed below the scarp which is here clearly the Interbasaltic Bed. The scarp feature is unbroken and climbs gently to the north-east and there is no indication of faulting. It is evident that, apart from the deep hollow at Craigahullier Quarry, the surface of the Interbasaltic Bed fell quite steeply from Ballycraig mine towards the quarry, the difference in level being about 18 m, and that the two lowest flows of the Tholeiitic Basalts were ponded against this barrier.

Above the feature formed by the third flow at Craigahullier there is a further scarp formed by a fourth tholeiitic basalt lava. Seen only in small exposures, the total thickness of this flow cannot be directly measured but it is probably at least 15 m. Small crag and quarry exposures south-east of the big quarry show irregular columns of a slightly coarser rock than seen in the lower flows. It is very similar to that of the topmost flow seen in Islandmore Upper, 800 m to the south-west.

Beyond Ballycraig Mine the outcrop of the tholeiitic basalts sweeps round the Upper Basalt scarp of Scudion Craig [C 893 388], and is marked at its northern edge by the mines in the Interbasaltic Bed. North of Scudion Craig, ill-defined scarps suggest that only two flows are present, but it is possible that at Crocknamold [C 894 396] there are three. A large quarry in the lowest flow beside the road shows over 15 m of starchy thin irregular columns, sometimes showing radial grouping, with a few more massive columns at the bottom of the face in the south-east corner. A second flow crosses the secondary road with a small scarp 460 m north of Scudion Craig, and 90 m south of this an extensive but shallow quarry shows over 6 m of thin irregularly columnar lava resting on the reddened vesicular top of the second flow. It is probable that these three flows, in all not more than 37 m thick, persist down the hillside towards Ballyhome Lower. A roadside quarry [C 904 383], 275 m north-east of Ballyhome, shows over 6 m of thin starchy columns.

Across Islandbraddagh and Glentask the outcrop of the Tholeiitic Basalts is extensive but exposures are limited and only rarely more than isolated knolls. A small quarry [C 914 395] 730 m north-west of Ballytober shows 2.4 m of columnar basalt which is said to overlie lignite, not now seen. This flow must be the highest in the series. About 1000 m further north-west a roadside quarry [C 908 403] shows 9 m of very irregular columns over a 2.2 m colonnade. This rests on 1 m of kaolinised vesicular tholeiitic basalt, presumably the top of the lowest flow. It is probable that there are at least four flows in the succession in this district.

North of Glentask Mine [C 909 407] the two lowest lavas form clear scarp features. Three metres of irregularly jointed basalt overlies the Interbasaltic at the north level of the mine, but at Boneyclassagh Farm a small quarry [C 914 414] shows 3 m of regular columns passing up into 5.5 m of irregular columns at the bottom of the first flow. The second flow is seen as crags of very massive basalt rarely showing any sign of columnar structure, but with flow-banding etched out on weathered surfaces.

At Ailsacraig [C 916 415] the first flow is seen in a down-faulted block beside the main road and a quarry shows 3.7 m of fairly regular columns under 12 m of irregularly jointed non-columnar basalt. At the top of the face this passes into 7.3 m of irregular columns. It is possible that this thickness, which is only part of the whole flow, is due, like Craigahullier, to infilling of a surface irregularity in the Interbasaltic Bed.

Good trap featuring is seen in the area about 1.6 km west of Bushmills where flow scarps emerge from the drift cover, but is much less conspicuous to the east. In the River Bush there are extensive exposures of basalt at The Walk Mills [C 939 398] and for some distance downstream. The scarp formed by the first flow is prominent west of the town, and, stepped north by a fault, on the west of the river downstream from the upper bridge.

East of the River Bush exposures are scattered and usually reveal only small thicknesses of individual flows among the peat bogs. A small quarry [C 954 407] south-east of Dundarave shows 4.6 m of massive, crudely columnar basalt and locally there are small flow scarps in Carnkirk and Lisnagunogue. Much more exposure is seen on the high ground north of Croaghmore, where strong trap-featuring above the outcrop of the Interbasaltic Bed appears to show six flows of the Tholeiitic Basalts. A small adit has been driven in the thin lateritic flow top at the base of the scarp formed by Flow 3, about 1.6 km north of Croaghmore. This deeply weathered top of Flow 2 is seen again at Craigalappen [D 024 415] where up to 2.3 m of ferruginous lithomarge is exposed below 6 m of columnar basalt. A small quarry in the lowest flow at Ballynastraid [D 019 430] shows nearly 4 m of starchy columns above 5.5 m of regular colonnade and is a useful source of good specimens of columnar rock. Several smaller quarries in Flow 2, a few hundred metres to the south also show similar material.

East of Croaghmore the width of the tholeiitic basalt outcrop reaches about 7 km though the thickness and number of flows have probably remained the same. A strong feature formed by the third flow is seen at Black Park [D 035 401] but south and east of this drift cover and extensive bogland make exposures few.

On Lanninmore Hill, south of Ballintoy, an extensive quarry in the lowest flow shows 6 m of columnar basalt which was once worked to make paving setts. The second flow forms an extensive outlier to the east, and, with the third, occurs at the top of the Hill. Higher flows are overlain by the peat bog but form small features which emerge through the peat at The Guy Stone.

Upper Interbasaltic Bed

The western outcrop of this bed extends for about 1.6 km northwards from the point where it intersects and is cut out by the Portrush Fault, just north of Cloyfin [C 880 353]. The position of the outcrop is clearly fixed by a prominent scarp of the Upper Basalts but the Interbasaltic rocks are exposed only in old mine workings at Killygreen and South Ballylagan. At the former [C 880 361] two adits driven in the base of the overlying lava are partly open but no section of the lateritic bed is visible. Red and yellow-brown ferruginous bauxite debris is seen on the tip and there is a strong flow of water from the southern adit. These mines were described in the 1888 Memoir as 'lately opened' in rich iron ore though bauxite was said to have been worked there previously and then abandoned (Symes and others, 1888, p. 24). On the field map, however, Symes stated that in 1886 the iron ore had proved a failure but that bauxite 1.2 m thick was being worked. The bauxite was described as reddish-brown, yellow, cobalt-blue, and violet in colour and lignite was noted in the spoil. Cole (1912, p. 20) gives a section recorded by the mine manager as:

Section in Killygreen Mine
Lignite	0.02 m
Hematite	0.1–0.3 m
Red Clay	0.61 m
Pavement	6.1 m
Lithomarge	1.8 m *base not seen*

'Pavement' was the term used by the miners for the ferruginous bauxite or lateritic lithomarge which underlies the iron-rich *cuirasse-de-fer*. Cole also remarks on the occurrence, apparently in the 'pavement', of fossil wood pseudomorphed by hematite.

Four hundred metres north of the Killygreen mines, two adits are seen north-east of South Ballylagan Farm [C 878 367]. The northern had fallen in but in the southern the following section was seen in 1958:

Section at Ballylagan Mine
	m
Basalt, coarse, columnar	2.4
Basalt, vesicular	1.2
Laterite, heavy dark-red ferruginous, slightly pisolitic in uppermost 0.15 m	0.45
Bauxite, reddish-brown and yellow, ferruginous, with spheroids of unweathered porphyritic tholeiitic basalt	2.75+

These mines were marked on the 1886 field map but no indication is given of when they were worked. An output of 50 tons of bauxite from Ballylagan is recorded in 1884.

About 230 m north of Ballylagan mine 'ore and lithomarge' were formerly seen in the track which runs up the scarp from North Ballylagan Farm, but only laterite fragments in the drift are now seen. Beyond this point the outcrop follows the base of the upper scarp across Islandmore Upper and is nowhere exposed. A small trial pit at the south boundary of the townland was driven at the wrong horizon, below the lower scarp, and found no ore.

From Islandmore the outcrop of the Upper Interbasaltic Bed follows a sinuous course across country to the east, located only by the scattered outcrops of Tholeiitic and Upper Basalts and nowhere exposed, for upwards of 12 km.

Around the hill of Croaghmore, which is capped by a spur of Upper Basalts, the course of the bed can be followed for some distance. A small exposure on the north-west face of the hill shows lithomarge, and the old maps note lithomarge with a little 'pavement' (ferruginous bauxite) above it at this point.

East of Croaghmore the outcrop is again lost beneath drift and its course is indicated only by the distribution of Tholeiitic and Upper Basalts. On the eastern edge of the sheet bauxite is said to occur in drains at the margin of a peat bog about 820 m north-north-east of Moyarget Lodge [D 075 388]. No rock is now exposed, but red ferruginous bauxite was seen in the spoil from a well north of Black Knowe, 270 m east of the sheet boundary.

Upper Basalts

West of the River Bann the only exposure of rock is of vesicular basalt seen in the bed of the Articlave River 140 m south-west of Ardina Bridge [C 788 348]. Less than a metre of rock is seen below boulder clay.

In the east bank of the Bann, basalt is exposed at several points at and below high-water mark. Massive compact basalt is seen between 640 and 470 m downstream from Craiganariff Fort [C 823 351] and 550 m upstream from the Fort coarse-grained basalt is exposed for over 90 m. Rock is exposed in the tail race of Drumslade Corn Mill [C 829 350] and in the steep bank to the west. On the southern edge of the sheet coarse-grained basalt is seen in the bank above the alluvial flat in the townland of Magheraclay.

North of Craiganariff Fort a quarry, opened to provide material for the Bann Estuary moles, is in massive rudely columnar medium-grained basalt with a maximum thickness exposed of 7.3 m. The base of the flow is seen in the west face of the quarry where it is slightly undulose and rests on vesicular lithomarged basalt. In the south-east corner of the workings the lowest metre of the massive basalt is vesicular and it rests on 15 cm of red laterite over 60 cm+ of vesicular kaolinised basalt.

In the townland of West Crossreagh there are a number of exposures of coarse-grained basalt in the bank above the raised-beach flat and near the margins of the blown sand, south-west and north of Dooey. Just 60 m north-east of Dooey an abandoned quarry in Glebe [C 822 364] exposes 3.6 m of massive porphyritic basalt.

The thin drift cover permits fairly extensive exposures of basalt in the townlands of Cappagh More, Cappagh Beg, West Quarter and Kiltinny More, in many of which small quarries have been opened. Most of these are long abandoned and often filled in but one [C 833 363], 365 m east-south-east from Cappagh More Farm, shows 2 m+ of coarse compact basalt with curious vertical bands about 30 cm wide of rotten rock containing small relict 'pebbles' of hard basalt. Further south a quarry [C 838 357] 460 m east-north-east of Ballygallin shows 2.4 m+ of coarse massive basalt and at Dooey 4.6 m+ of rubbly medium-grained basalt is seen in a quarry face. At Kiltinny Beg Farm [C 845 353] 3.6 m+ of massive medium-grained basalt is exposed in a small working.

East of the railway there are numerous small exposures of basalt, generally coarse-grained, in the townlands of Islandvardin, Bellemont and Roselick Beg, and similar exposures are seen west of the railway in Ballyleese. As usual only the hard parts of the lavas appear among the boulder clay as rock knolls and all the exposures are fresh and compact, but only in North Ballyleese has the rock been quarried and the workings [C 835 371], 275 m north-north-west of Ballyleese Farm, show over 3.6 m of massive medium-grained basalt.

Near the Islandmore Fault there are extensive exposures of coarse feldsparphyric basalt in East Ballymaclevennon, Inchmearing, and Craignahorn. Crags show up to 9 m of massive rudely columnar rock which has been quarried on a small scale at several places. All the exposures are probably on the outcrop of a single lava.

The greater part of the townlands of Maddybenny is drift-free and there are extensive exposures of basalt. Over 3 m of massive fine-grained rock, sparsely vesicular throughout, is seen in an old quarry [C 853 372] 365 m north-north-east of Island-tasserty cross-roads, and over 3.6 m of medium-grained basalt is exposed in a small quarry 30 m north-west of Maddybenny More Farm [C 850 376]. There are also extensive exposures on the high ground north of Craigtown More Farm, with over 6 m of massive feldsparphyric basalt seen in a small quarry [C 839 384] 30 m south-east of Craigtown Beg.

The high ground of Juniper Hill and East Ballygelagh displays crags of massive flow-banded medium-grained basalt, weathered faces often showing good etching of the flow line. Much-jointed coarse-grained basalt is seen in the railway cutting, and, in a small quarry south of the railway south-south-east of Port Gallen, blocky-jointed medium-grained basalt is seen to overlie the irregular lateritised top of a lower flow. The high ground of Ballyreagh is also largely drift-free and there are exposures of basalt in the railway cutting and along the minor roads near Ballyreagh House. Small quarries [C 847 393] 275 m east-north-east of the House show medium-grained basalt resting on the lateritised irregular top of a lower flow.

On the south-east outskirts of Portstewart a drift-free area in the middle of the blown sand has been the site of two quarries near the Old Coach Road and 365 to 460 m east of Queenora Cottage. Partially filled-in, one still shows over 3 m of spheroidal-weathering vesicular basalt [C 824 382]. Less than a kilometre to the east, crags of massive coarse-grained feldsparphyric basalt on a north–south scarp which crosses the Golf Links have been quarried at Rockview [C 831 384] where over 3 m of basalt is exposed in old workings.

The cliffs at Strandhead at the east end of Portstewart Strand are of massive rudely columnar basalt with horizontal flow banding seen at some places. The rock has a poikilitic texture clearly seen in hand specimens and the flow is at least 9 m thick. An exposure, at the base of the cliff west of Tubber Patrick [C 812 367], of thinly jointed fine-grained basalt may represent the base of the flow, but in the small bay north-west of the Strand Hotel the base is clearly seen resting on the lithomarged top of a lower flow. Though exposures are not continuous it is probable that the rocky foreshore as far as Rock House [C 813 370], and the cliffs behind the House, are all in this same flow, and the slight northerly dip towards The Berrins [C 812 375] suggests that the rocks on the platform almost as far as the sewage outfall are also part of the same lava. South of The Berrins is overlain by patches of a succeeding fine-grained flow. The flow-banded fine-grained basalt seen at the seaward end of the jetty at The Berrins beyond the small volcanic vent appears to be the representative of a lava not seen elsewhere.

Between The Berrins and Portahapple [C 813 377] the foreshore is underlain by three gently undulating flows of slaggy and vesicular basalt with lateritic tops, the thickest being some 3.6 m thick. North of Portahapple the dip steepens suddenly to 20° to 30° to the south, and the foreshore on Big Ringan and the cliff to the east show a series of about eight thin lavas.

Section east and north of Portahapple	m
Basalt, spheroidal-weathering, below soil	1.0+
Basalt, spheroidal-weathering with reddened top	2.4
Basalt, fairly massive, medium-grained, with slaggy top and base	3.6
Basalt, fine-grained, flow-banding shown by lines of vesicles	2.4
Lithomarge, purple, with relict spheroids of basalt and flecks of laterite at top	0.6–0.9
Lithomarge, purple, with lateritic band at top. Colour banding	0.6–0.9
Laterite, bright red, passing down into pink bauxitic lithomarge	1.5
Basalt, vesicular spheroidal-weathering	3.0
Basalt, massive, slaggy-topped, coarse	6.0+

Opposite Little Ringan the lowest flow of the Portahapple succession is overlain by the nose of a flow of coarse-grained massive basalt which thickens rapidly northwards to over 6 m and forms the foreshore as far as the Convent point. This lava is very vesicular in the uppermost 2 to 2.5 m and the top part of the compact basalt has horizontal flaggy jointing. West of the Dominican Convent the dip steepens to about 20° to the north-east, forming the northern limb of the anticline with axis about Little Ringan, and on the foreshore round the Convent point and in front of The Crescent a series of flows is exposed which are the equivalent of the Portahapple succession. Because of the impersistent and irregular nature of these flows, however, no exact correlation is possible.

Section on foreshore opposite The Crescent   m

Basalt, massive medium-grained seen as outlying 'islands' north and east of swimming pool and as seaward rocks to the west	3.0+
Basalt, fine-grained vesicular. Flow-banding picked out by compact non-vesicular bands. Irregular lateritic top	~2.0
Basalt, compact, with clean-etched flow-banding on weathered surfaces. Reddened top	2–3
Basalt, very vesicular (in bands). Thins out to north	0–2
Basalt, massive, medium-grained, flaggy jointing in uppermost 2 m. Thins to north	6–9
Basalt, massive, coarse-grained. Thins to south	6.0

The foreshore west of Main Street, Portstewart, shows three, or possibly four, flows of basalt. The lowest, a slaggy vesicular spheroidal-weathering basalt is probably the same as that seen at the swimming pool, and its base is seen just north of the Town Hall where red laterite is seen just below the sea wall. East of the Isle of Benoney, outlying patches of a higher flow are seen and 90 m south of the harbour the base of this flow overlies the slaggy reddened top of the lowest lava. The newer flow is about 6 m thick and is overlain just south of the harbour by 2 m of massive, medium-grained basalt. On the landward side of Main Street an old quarry in the raised-beach cliff which backs the bay shows two flows dipping south-westwards which may be the lowest two of the foreshore succession.

Portstewart Point is formed by a very massive fine-grained basalt lava over 14 m thick, well seen in the cliffs and the old quarry at Kenora Terrace [C 815 387]. The base of this flow is seen at various points round the headland and the deep gullies north of the harbour and at the Herring Pond have been formed by erosion along the lateritised and lithomarged top of the underlying lavas. The strong vertical jointing, often orientated north–south, gives much of the cliff section a rudely columnar appearance. At a point about 67 m south of the Point the jointing is concentric, circular, and vertical, while 135 m east-south-east of the Herring Pond the joints form a large pitching 'anticline' on the foreshore. This is near the bottom of the flow and may be a 'magmatic roll'. The rock exposed in an old quarry on the north side of the Old Coach Road at Harryville [C 818 385] is also fine-grained basalt, sparsely porphyritic, and may be part of the same flow.

At Atlantic Circle the base of the Point flow is seen to rest on a thin flow of rubbly basalt about 3 m thick, and, with the dip swinging to the east, it appears again in about 90 m and dips below sea level. Both above and below this flow here there are thin rubbly lavas with red lateritised tops, the lower of which is seen to wedge out on the foreshore. From here to Black Rock the foreshore for almost a kilometre shows blocks and rafts of medium and fine-grained basalt in a matrix of laterite and lithomarge, presumably the top of a deeply weathered lava. These outlying blocks are apparently relics of the overlying flow which is seen for about 180 m at the Blowing Hole where it reaches a thickness of 8 m. It appears to thin out to the north-east and is probably represented by the flow that borders the southern edge of Port Cool and is overlain by the massive rudely columnar coarse-grained lava, over 6 m thick, forming the Dog Rock [C 823 391]. This flow can also be followed along the southern side of Port Cool for about 275 m.

At the south side of Black Rock the Blowing Hole flow seems to be underlain by another thin kaolinised and stained lava which outcrops on the foreshore. Immediately below the Rock reddish lithomarge which may be the top of a further flow is seen at low water, but exposures here are not clear and it is probable that the Black Rock intrusion has disturbed the lavas in its vicinity.

Between Port Cool [C 825 391] and Stony Port [C 829 393] the promontory is a raised-beach platform cut in the top of a massive flow of medium-grained basalt with rude hexagonal jointing and some flaggy jointing. Small outliers of the overlying flow are seen on the platform. On the east side of Stony Port the flaggy-jointed coarse basalt seen in the raised-beach platform may be the same flow and here is overlain by a lenticular flow which reaches a thickness of 6 m and dies out again in a distance of about 60 m. This in turn is overlain by a massive lava 10 m or more thick which can be followed to the cliff top at Holywell Port, where it has two thin flows beneath it, both of which die out rapidly to the west. Beyond this point the foreshore and cliffs consist entirely of thin slaggy vesicular lavas, rarely over 3 m thick. Five or six of these flows can be counted and not until the west end of Port Gallen [C 835 395] is a massive lava seen, in this case a 9-m+ flow in the cliff.

East of Port Gallen the low cliff and rock platform are cut in an 8-m+ flow of coarse basalt, flaggy at the top, which is seen to rest on the slaggy and vesicular top of a lower flow. A gentle easterly dip brings in another massive coarse basalt lava in the cliff 90 m east of Port Gallen and this flow forms the foreshore as far as the small fault beyond the two small volcanic vents described below as the Craigtown vents. On the seaward side of the vents a 15-cm vesicular band in the lava contains a block of sandstone 23 cm across and on the landward side the lateritic weathered top of the same lava contains blocks of pale-brown recrystallised limestone. The proximity of the explosive vents to these occurrences is notable, but there is no obvious connection between the xenoliths and the intrusions.

Beyond the small fault the foreshore shows a series of three thin basalt flows for 180 m when a second fault cuts this succession off against 3 m of slaggy basalt overlain by 6 m of massive medium-grained basalt. This latter flow forms the foreshore for over 275 m to Long Port [C 842 398] and the succession seen in the cliffs behind Sycamore Port [C 840 398] is:

Cliff section east of Sycamore Port   m

Basalt, massive, thinning out and disappearing eastwards	9+ to 0
Basalt, coarse-grained, vesicular in topmost 3–6 m	9.0
Basalt, rubbly, coarse, irregular, vesicular top	3.6
Basalt, massive, medium-grained on foreshore	6.0+

South of Long Port these flows pass laterally into a jumbled mass of basalt blocks and laterite slabs. This is not like the normal explosive vents seen on this coast and is more probably a fault or collapse feature.

East of Long Port the cliffs and foreshore display a series of lavas dipping gently eastward:

Sections between Long Port and Devil's Port   m

Basalt, massive, close-jointed	3.0+
Basalt, massive, lithomarge band 1 m thick at top	12.0
Basalt in two or three thin irregular and lateritised flows	9.0
Basalt, massive, coarse-grained	5.5
Basalt, massive, very fine-grained, vesicular reddened top	9.0
Basalt, massive, coarse-grained, vesicular in topmost 3 m	7.6
Basalt, medium-grained, with ~0.3 m lateritised top	3.0+

Devil's Port and Old Castle Port [C 845 339] are controlled by faults and a third fault runs along the east side of the spur below Ballyreagh Castle. The western sector, behind Devil's Port, is largely occupied by a volcanic vent, but two flows of fairly massive basalt are seen in the cliffs. Immediately west of the Old Castle Port Fault the slaggy top of the lower flow contains a 15-cm block of recrystallised brown limestone, confirming the impression at Craigtown that some explosive activity took place in this area during the extrusion of the lavas.

The cliff below Ballyreagh Castle and the spur beyond it show about 18 m of a composite flow with about seven bands of compact spheroidal-weathering basalt, and averaging about 1.5 m in thickness, interbedded with beds of grey slaggy vesicular basalt, some slightly reddened. There is a conspicuous red laterite bed at the top of the banded series, and the whole is overlain and underlain by massive basalt lavas, only a metre or so of which is seen. This banded flow is similar to that seen at Dunluce (p. 116) and also to the composite flow which occurs near the top of the Lower Basalts on Rathlin Island.

Between the fault at Ballyreagh Castle and that at Black Hill Port [C 848 399] the cliff sections and raised-beach rock-platform, affected by two minor faults, display four flows of basalt:

Cliff section west of Black Hill Port	m
Basalt, much-jointed, fine-grained	9.0+
Basalt, massive, medium-grained; top 1.5 m vesicular but not reddened	6.0
Basalt, massive, coarse-grained	6.0
Basalt, massive, medium–fine-grained, forming rock platform	4.5

East of the Black Hill Port fault, which is of uncertain throw, a new series of lavas is exposed in the cliffs, the upper 9 m of which is probably the lower portion of the composite flow at Ballyreagh Castle:

Cliff section east of Black Hill Port	m
Basalt, massive, coarse-grained; very weathered with veins of more compact material	3.0+
Basalt, slaggy, vesicular, with irregular compact bands	0.9
Basalt, coarse, irregularly jointed, sparsely vesicular	3.0
Basalt, flaggy-jointed with small vesicles	2.4
Basalt, massive, fine-grained; conspicuous reddened top	9.0
Basalt, massive, medium-grained; vesicular in upper part; 0.6–0.9-m band of purple lithomarge at top	6.0–7.6
Basalt, deeply weathered, rubbly-weathering, with thick vesicular top and 0.6–0.9-m laterite bed with blocks of rotten basalt	6.0+

At Black Hill Cove a fault hading east has been eroded to form a cave. To the east of this the cliffs and rock platform display a gentle anticline in a series of lava flows, the lowest of which is probably the upper part of the banded flow, indicating a downthrow of the order of 24 m.

Section seen east of Black Hill Cave	m
Basalt, massive	2.0+
Basalt, massive, irregularly jointed; conspicuous red top	4.5
Basalt, massive, irregularly jointed; vesicular top	7.6
Basalt, very vesicular, spheroidal-weathered with reddened top thinning out westwards	6 to 0
Banded or composite flow. Lenses and bands of compact basalt interbedded with grey vesicular basalt. Some faint reddish coloration on some vesicular beds.	9.0+

About 365 m east of Black Hill Cave the rock on the foreshore disappears beneath beach boulders and sand and no further exposures are seen.

Area east of Portrush Fault

Where the Upper Basalts overlie the Upper Interbasaltic Bed in the Islandmore–Ballylagan area there are no extensive exposures of the lowest lavas but the basal flow seems to be rather thin. It is seen over the old mine entrances at Ballylagan [C 878 367] where it is coarse-grained and rudely columnar but only a metre or so is exposed. Over the old Killygreen mines [C 880 362] up to 2.4 m of decomposed vesicular medium-grained basalt is seen. Excavations for a small covered reservoir in North Ballylagan [C 876 370] showed over 3 m of slaggy vesicular basalt with spheroids of fresh medium-grained basalt, which must be near the base of the series. About 75 m further north a small quarry shows 4.6 m of massive blocky-jointed basalt overlying the irregular slaggy top of a medium-grained lava of which only 1 m is exposed. The base of this exposure is some 4.6 m above the terrace which marks the top of the uppermost tholeiitic lava and the concealed section must include the Upper Interbasaltic bed, thin here, and the basal flow of the Upper Basalts. Though the lowest flow forms a distinct scarp feature north and east through Islandmore, other small exposures indicate that individual flows are thin, vesicular bands being common and massive compact lavas not being seen to exceed 3 m in thickness.

A few small exposures of coarse basalt are seen on the slight scarp feature which marks the base of the Upper Basalts across the southern edge of Corbally townland. No further outcrop is seen across over a kilometre of country, but at Scudion Craig [C 893 389], a conspicuous landmark, the lowest flow seen of the Upper Basalts, is over 6 m thick and consists of massive rudely columnar fine-grained basalt with some horizontal flow banding on weather-etched surfaces. The base of this flow is obscured by scree and the lowest rock seen is some 4.6 m above the surrounding tholeiitic basalts, but it is likely that the flow is massive for this thickness and is not underlain by thinner lavas. The lavas dip south-east at 3° and the trap feature formed can be followed for about a kilometre across Ballyhome. Succeeding flows form thin and discontinuous features and a quarry 460 m south-east from Scudion Craig shows parts of three thin flows in a thickness of only 5 m. Numerous exposures and small quarries in these thin lavas are seen in the area south and west of Ballyhome Lower.

A very massive flow of fine-grained porphyritic lava is seen in quarries on both sides of the minor road about 1000 m due north of West Park from where up to 7.6 m of basalt is exposed. This flow is apparently near the base of the series but exposures are very scarce in the area eastwards towards Ballytober and the exact position of the boundary between the Upper and Tholeiitic basalts is obscure.

In the southern part of the townlands of East and West Park a well defined drift-free plateau is formed by a massive flow of olivine-phyric fine-grained basalt which has been worked in several quarries. The largest, just east of the main Ballymoney–Portrush road [C 907 376], shows 7.6 m of massive rudely columnar fine-grained chrysophyric basalt resting on the irregular slaggy top of a lower flow. Other quarries near the Bushmills–Coleraine road show over 6 m of the same flow.

Over a kilometre to the west the Crag of Dunmull [C 889 371] stands up as a flat-topped outlier of a similar olivine-phyric flow, possibly the same one. The upper part of the Dunmull 'plateau' is of massive basalt, irregularly jointed at the top and rudely columnar below, some 6 m high, with its base concealed by talus. The terrace to the south-west of the 'Dun' is formed by the upper part of a lower rudely columnar coarse-grained flow.

South of the Coleraine–Bushmills road exposures are very scarce, the only important one being on Out Hill [C 901 365] behind Ballywath Manse, where an old quarry shows over 3 m of massive fine-grained chrysophyric basalt, and a similar rock is seen in small exposures on the low ground to the east. It is not improbable that Dunmull, East and West Park, and Out Hill are all outliers of a single massive lava flow. There are small exposures in the Flower Hill and Island Carragh area but they never show more than about a metre of olivine-basalt.

East of the River Bush there is a large quarry near Seneirl Bridge [C 942 360] where a 15-m face shows massive rudely columnar basalt. Scattered small drift-free areas between Islandahoe and Billy have a few small exposures, but more extensive exposures occur around Craig Hill [C 964 390], where a roadstone quarry showed about 5 m of massive olivine-basalt with the amygdaloidal top of a lower flow in the quarry floor. About 460 m north-east from Turfahun 3 m of olivine-basalt is seen to overlie 2 m of tholeiitic basalt in the side of a lane. The Upper Interbasaltic Bed here is represented by only a few centimetres of inter-lava material.

Though there are fairly extensive drift-free areas around Liscolman only rare exposures of rock are seen. More revealing is the Croaghmore area, where features on the hill indicate the presence of three flows, the upper two as outliers. A section of about 4 m of poorly columnar olivine-basalt is seen on the west side of the hill in the lower part of the second flow. An old quarry just south-west of Croaghbeg crossroads shows 5 m of weathered amygdaloidal basalt, probably near the base of the lowest flow; and many of the other small exposures in the Ballyoglagh area are probably in the same flow, including the quarry [D 000 380] where up to 5.5 m of massive basalt is overlain by up to 2 m of reddened ferruginous laterite.

Along the drift-free ridge to the south four later flows can be distinguished, with small quarry and crag exposures, all of massive olivine-basalts.

The eastern part of the outcrop is very poorly exposed, the only extensive drift-free area being around the village of Moyarget. Here 4.5 m of massive poorly-jointed olivine-basalt was exposed in a quarry near the Masonic Hall. Similar rocks are also seen around Sproule's Town and in a small quarry on a knoll in the middle of the bog [D 062 364] where 5 m are seen.

North of Portbraddan Fault there are two outliers of the Upper Basalt. The larger, in the townlands of Carrowreagh and Dunseverick is exposed in small drift-free knolls south of Tonduff and in small quarries some 500 m west of Feigh Upper where thicknesses of 2.4 m and 1.5 m of olivine-basalt are seen. East of Feigh small exposures occur in streams.

The smaller outlier lies west of Portbraddan where the cliffs behind Portacallen show several thin flows of olivine-basalt, usually very vesicular. In one there is some development of basalt pegmatite, associated with large vesicles. H E W

PETROLOGY OF THE EXTRUSIVE IGNEOUS ROCKS

A large number of thin sections have been examined, mainly by J. R. Hawkes and D. C. Knill, but a few by R. W. Elliott and H. E. Wilson.

Lower and Upper Basalts

Both groups consist of olivine-basalts which show considerable local variation, even in the same flow, but which, as groups, cannot be distinguished. Petrologically they are holocrystalline rocks, consisting essentially of labradorite feldspar, olivine, augite and iron ore, with interstitial chlorite and zeolites in varying proportions. Zeolites and, rarely, chalcedony, occur as vesicle fillings.

Both olivine and feldspar occur in two generations—as small grains in the groundmass and as phenocrysts. Augite occurs as ophitic or sub-ophitic plates, and occasionally as a granular constituent of the groundmass also (NI 1487).

Modal analyses of a number of flows of the Lower Basalts, collected from the area to the east of the sheet, indicate that the proportions of the constituents vary considerably, not only from flow to flow but also from different localities in the same flow.

The most abundant constituent is plagioclase feldspar which generally forms just over half the rocks by volume. Most of the specimens show only labradorite but zoning to andesine was recorded in one case (NI 1555). Though laths are usually random in orientation there is occasional fluxion-texture (NI 1538).

Feldspar phenocrysts are fairly common and are also of labradorite. Twinned crystals up to 2 mm (NI 1387, 1403) are recorded and, as in the groundmass, the grains usually show some alteration to chlorite. Interpenetrating glomeroporphyritic aggregates occur rarely (NI 1456).

The pyroxene is a pale lilac or colourless augite, often titaniferous, which forms 15 to 30 per cent of the rock by volume. It occurs almost invariably as ophitic plates, up to 3 mm across and commonly showing idiomorphic form, which enclose feldspar laths, olivine grains, and magnetite crystals. In some cases the augite plates form a complete mosaic but usually they are sub-ophitic and there is a certain amount of interstitial material. Locally there has been some replacement of the pyroxene by chlorite (NI 1484).

Olivine, generally more or less altered, makes up 10–15 per cent of the rock by volume. It frequently occurs in two generations—macrophenocrysts up to 3 mm in diameter (NI 1387, 1396), which may tend to be glomeritic (NI 1485), and smaller grains (0.1–0.2 mm) which occur in the groundmass and as inclusions in ophitic pyroxene. The phenocrysts are not always of the same composition as the groundmass olivines, the latter having higher refractive index (NI 1396).

Though sometimes fairly fresh the olivine grains always show some degree of alteration to bowlingite, iddingsite, chlorophaeite, chlorite or clay-mineral, and often the original material has been completely replaced.

Iron oxide—probably magnetite—is ubiquitous, occurring in rods and idiomorphic grains, as inclusions in olivine phenocrysts and as a component of the groundmass. Its abundance varies from the usual 1 or 2 per cent to 10 per cent or more in rare cases.

Zeolites occur both as vesicle fillings and as original constituents of the rock, where they form interstitial patches and represent the final products of crystallisation. Chabazite, natrolite and analcite are recorded as vesicle fillings—with chalcedony in one case—and analcite, chabazite, natrolite, phillipsite, thomsonite and stilbite occur in the mesostasis. Carbonate, probably calcite, also occurs sporadically in both vesicles and groundmass.

Chlorite, varying in colour from dark to pale green, and occasionally brown, occurs in virtually every specimen, usually as part of the groundmass, locally replacing feldspar or, less commonly, augite, and also occurs, rarely, as a vesicle-filling. This mineral may make up to 20 per cent of the rock by volume.

One specimen (NI 1391), which is from part of a banded pahoehoe-type lava, contains small patches of dolerite-pegmatite of the same composition as the rest of the rock—labradorite, augite, olivine—but with more mesostasis.

Chemical analysis of the olivine-basalts gives the following results (average of seven analyses, Patterson, 1952; Tomkeieff, 1940):

SiO_2	45.64	per cent
Al_2O_3	14.77	
Fe_2O_3	2.87	
FeO	8.29	
MgO	11.29	
CaO	10.06	
Na_2O	1.97	
K_2O	0.28	
TiO_2	1.31	
P_2O_5	0.2	

Tholeiitic Basalts

These rocks differ from those of the Upper and Lower Basalts in several ways. In the field they can be recognised by extreme fineness of grain, a tendency to conchoidal fracture and the absence of vesicular bands except at the extreme bottom and top of flows, which are commonly thick and persistent.

Petrologically the tholeiitic basalts are hypocrystalline rocks with an abundant mesostasis of glass, often devitrified or chloritised. Feldspar occurs as minute laths and sparse phenocrysts; pyroxene is present as small grains, often idiomorphic and never ophitic. Olivine is rare though pseudomorphs after it are fairly common. The scarcity of olivine is matched by a poverty in magnesia, while the silica content is about 5 per cent higher than that of the olivine-basalts.

The typical tholeiitic basalt consists of a mat of feldspar laths and granular pyroxene with patches of mesostasis and scattered grains of magnetite. The extremely fine grain is reflected in the average linear intercept of mineral crystals which is about 0.03 mm. Most of the specimens are microporphyritic, usually with scattered feldspar phenocrysts but occasionally with augite or olivine.

Plagioclase feldspar occurs in two generations. Sparse microphenocrysts and rare porphyritic crystals up to 3 mm long, commonly euhedral, are known. The microphenocrysts are generally about 0.5 mm long with some inclusions of pyroxene. The larger euhedral grains are occasionally glomeritic and are slightly more calcic than the labradorite of the groundmass (NI 1395, 1567, 1570).

The groundmass feldspar is in the form of small laths averaging about 0.1 mm in length and 0.02 mm in width though some specimens are appreciably coarser in grain. The feldspar is labradorite (An_{50} to An_{55}). Fluxion texture is rare and the laths are usually random in orientation, but specimens from the uppermost flow of the series near Portbraddan show good flow alignment (NI 1568–71) and a specimen from the third flow from the top displays partial alignment (NI 1562).

Pyroxene occurs as tiny tabular and granular crystals, sometimes showing idiomorphic form. The granules are generally packed in small tight clusters but also occur as isolated grains. Sub-ophitic texture, with grains enclosing feldspar laths, is rare. The mineral is usually sub-calcic augite ($2V=35°–40°$) but occasional 2V measurements suggest compositional variations towards pigeonite. It appears, however, that pigeonite is uncommon and not the dominant constituent as implied in the Ballycastle Memoir (Wilson and Robbie, 1966, p. 205).

Pseudomorphs of serpentine after olivine are common but these scattered and very small grains usually make up only one or two per cent of the rock, though exceptionally it can amount to 5 per cent. Fresh olivine has been recorded from the sixth flow of the series at Portnabrock (Patterson, 1955, p. 104) and from the second flow on Rathlin Island (Wilson and Robbie, 1966) but has not been seen in specimens from the present collection.

Iron oxide is abundant in the form of small grains in the groundmass, usually accompanying pyroxene crystals.

The non-crystalline mesostasis is partially glassy but it is chloritised in places or full of opaque granular material. The glass is generally devitrified and varies in colour from brown to green, commonly near opaque from its load of iron oxide dust. Tomkeieff (1940, p. 109) describes the mesostasis as 30 per cent brownish glass, 50 per cent chlorite—mainly palagonite with some isotropic chlorophaeite—and 20 per cent iron oxides and other minerals.

Chemical analysis of the tholeiitic basalts gives the following results (average of ten analyses, Patterson, 1955):

SiO_2	50.65	per cent
Al_2O_3	14.28	
Fe_2O_3	2.95	
FeO	7.64	
MgO	5.15	
CaO	10.4	
Na_2O	2.71	
K_2O	0.76	
TiO_2	0.99	
P_2O_5	0.46	

Pyroclastic rocks

The bedded tuffs associated with the Carrickarade vent are by no means homogeneous and there is a considerable variation in the composition of the material from place to place. The main macroscopic constituent of the tuff is chalk, usually comminuted but locally coarse enough to be described as agglomerate. There are also macroscopic fragments of flint, basalt, Liassic mudstone and, very rarely, sandstone. Blocks of an earlier pale-brown tuff have also been recovered (Patterson, 1948, p. 207).

Thin sections of the matrix show it to consist of lava fragments, glass and granular calcite. The lava fragments are largely of feldspar laths set in a glassy matrix (NI 1518–20), sometimes rich in zeolite. The glass is olive-green in colour and consists of shards and also of pumice. Calcite is an interstitial constituent and also occurs infilling bubbles in the glassy pumice. Organic material —foraminiferal tests—can be distinguished in the matrix. In the lowest 2 metres or so of the tuff there is heavy discoloration by manganese deposits. Patterson records up to 90 per cent Mn in a sample from the stream near Ballintoy, and a thin section of this material shows heavy impregnation of the matrix, which is largely glass with scattered anhedral quartz crystals (NI 1521).

The occurrence of a bed of tuff above the first flow of the Tholeiitic Basalts on Rathlin Island is known (Wilson and Robbie, 1966, p. 201). A specimen of the inter-lava laterite from the top of the first flow at Craigahullier quarry revealed that the lateritised material here also is a glassy ash, consisting of rounded and elongated fragments averaging 0.4 mm of glassy material, devitrified lava, and basalt which is altered to gibbsite and iron oxides (NI 2113).

Similar thin lateritised tuffs probably occur at other levels. A laterite bed at Old Castle Port in the Upper Basalts contains a block of argillaceous limestone and must have been a pyroclastic deposit (NI 1398).

Interbasaltic Beds

The composition and origin of the interbasaltic laterites and bauxites were discussed by Eyles (1952) and his conclusions have general application. The following brief account of the processes of lateritisation is synthesised from Eyles' work:

The interbasaltic aluminous laterites of Antrim are of two main types: highly ferruginous red bauxites, formed from basalt, with 20 per cent to 30 per cent of iron oxide and less than 10 per cent of silica; and siliceous grey bauxites formed from rhyolitic debris, containing less than 10 per cent iron oxides and up to 40 per cent silica. The latter type has been recorded from Clegnagh Mine, Ballintoy.

The formation of the laterites was accomplished in two stages: first, alteration to lithomarge, a rock consisting mainly of hydrated silicate of aluminium—that is kaolinite or halloysite; and then the formation of laterite from lithomarge. Hydration is an essential part of both processes and is brought about by the circulation of aqueous solutions within the parent rock.

Lithomarge is formed by the breaking down of the complex silicates which make up basalt (feldspar, pyroxenes, and olivine) into their constituent oxides. Some of these, notably magnesia, lime, and the alkalis are eliminated in solution, with part of the silica, the remaining silica and alumina combining with water to form kaolinite or halloysite. The oxides of iron and titanium remain but most of the iron is oxidised to the ferric state. Lithomarge formed from basalt is characteristically purple or reddish purple in colour, spotted with segregations of paler kaolinite or halloysite. It is very common at the top of many lava flows in the Lower Basalts below the thin red laterites, and in the flows below the main Interbasaltic Bed the kaolinisation process has proceeded to considerable depth giving lithomarge 12 m or more in thickness.

Lateritisation is essentially a desilicification process, the hydrated silicates of alumina losing their silica in solution and the residual alumina taking up water to form the trihydrate of alumina, gibbsite ($Al_2O_3.3H_2O$). Concurrently, segregative and concretionary processes tend to effect a separation of the aluminous from the ferruginous oxide so that ultimately the iron oxide forms an upper layer of highly ferruginous laterite overlying a bed of aluminous laterite or bauxite. The upper layer, when sufficiently rich, constitutes an iron ore.

The process is a slow one and has rarely, in Antrim, been taken to completion before the onset of the succeeding lava. As a result, rocks of intermediate types—lateritic lithomarges and ferruginous laterites—are here the prevailing varieties.

The examination of specimens from the Interbasaltic Beds and from the inter-lava laterites in this area has revealed some new facts about these residual materials.

The lateritised top of the tuff bed at the Giant's Cut is a banded reddish-brown material which consists of irregular bands composed largely of hematite and goethite (NI 730–1). These ferruginous bands show highly irregular structure, but occasional basaltic textures can be discerned. X-ray examination of the clay mineral by B. R. Young shows it to be a montmorillinoid, probably montmorillonite, with some impurity, probably apatite.

The occurrence of montmorillonite is unusual in the Antrim laterites as the common clay minerals are kaolinite and halloysite. The cause of this anomaly is not certain but may be local weather conditions during formation, which affect the order of breakdown of the minerals in the parent rock. Simultaneous decomposition of the ferromagnesian minerals and the feldspars, with release of ferrous and ferric iron, magnesium, alumina and silica, leads to the formation of montmorillonite and beidellite. With early breakdown of the ferromagnesian minerals, however, magnesium and iron may be removed in solution before the feldspars decompose, with subsequent formation of kaolinite. Kaolinite, therefore, tends to form under conditions of prolonged weathering, while montmorillonite results where drainage is poor and oxidisation restricted.

Montmorillonite has also been detected by X-ray examination in bauxites from Ballylagan—the Upper Interbasaltic Bed—where a mottled bauxite proved to be composed of gibbsite and hematite in the red matrix and gibbsite and montmorillonite in the pale patches where the latter mineral is a small proportion of the whole (NI 2111). A specimen from the main Interbasaltic Bed at Lemnagh Beg also contains 1 to 2 per cent of a montmorillonite-group mineral in a matrix of disordered kaolinite and metahalloysite (NI 1525).

The mineral boehmite, previously recorded mainly from specimens collected close to basalt dykes and assumed to be caused by dehydration by heat, has been found in a lithomarge from Port Moon (NI 1539). It may have resulted from the effects of heat from the overlying basalt lava.

The occurrence of a relict mass of tholeiitic basalt in the Upper Interbasaltic Bed at Ballylagan was one of the arguments used to prove the existence of two Interbasaltic horizons in this area. This basalt spheroid (NI 210) is a porphyritic tholeiite with automorphic phenocrysts of anorthite, partially replaced by chlorite in the inner zones, in a groundmass of labradorite laths, granular augite, and a glassy mesostasis.

H E W

CHAPTER 17
Intrusive igneous rocks

VOLCANIC VENTS

THE CARRICKARADE VENT

The vent agglomerate

The vent agglomerate as exposed at present occurs on the mainland between Portaneevey and Larry Bane Bay and on the south-west side of Carrickarade Island. It is particularly well exposed at the rope bridge where a vertical thickness of 30 m or so is seen—much of it inaccessible. The agglomerate in the vent is of a darker basaltic type, consisting essentially of rounded blocks of basalt, some up to 3 m across, in a tuffaceous matrix of the same rock. The basalt blocks are always highly decomposed, probably through hydrothermal action rather than subaerial weathering (Tomkeieff and Patterson, 1948, p. 204). There are scattered pieces of chalk and flint.

On the western margin of the vent up to 20 m of dark agglomerate or tuff with basalt, chalk and flint, the basalt blocks ranging up to 0.6 m in diameter, can be seen above Casnageeragh, lying on Upper White Limestone. The grassed gully immediately to the east gives no hint of the exact relationship between this tuff and the intrusive dolerite forming Stackahorlin. The tuff thins rapidly to the west and near the postulated margin of the vent only 6 m can be observed beneath 3 m of drift. At the rope-bridge (which is 27 m above sea level) sections on the island are about 30 m high, and on the mainland agglomerate forms the cliffs for 100 m to the south-east and is seen to heights of 60 m north of Binard.

Pockets of dark agglomerate are seen at the bottom of the cliffs east of Binard and tuff forms the high bluffs west of Portaneevey which are locally veneered on their lower faces by dolerite. The dark tuff can be found 90 m or so above the sea near the vent wall where it is overlain by chalky bedded tuff.

Apart from a small occurrence near the tourist path about 80 m north-west of the observation point there are no inland exposures of the vent agglomerate, but it is conjectured to underlie the area north of the break in slope at the base of the bedded chalky tuff.

Vent intrusions

A number of intrusive masses of basalt penetrate the pyroclastic filling of the vent. In the west the intrusive mass of Stackahorlin is at or near the vent margin. Basalt exposures continue eastwards along the accessible part of the shore while vent agglomerate occurs in the cliffs above. On Carrickarade Island a small knoll a few metres north of the rope bridge consists of intrusive basalt. The lower transgressive contact dips south-east at 20°. About 4 m of this sill-like intrusion is seen; the basal 0.4 m contains a number of xenoliths of grey porcellanite and, rarely, baked chalk. The basalt itself is amygdaloidal and non-columnar. The larger mass of intrusive basalt forms the sea-stack to the north of the island and continues south to form the cliffs on the main island. The base of this sheet on the western side of the island dips at about 20° to the north. The sheet is markedly columnar.

On the mainland north of Binard an intrusive dolerite mass has a vertical contact with agglomerate to a height of some 80 m and another intrusion forms a line of crags on the west side of the gully along the contact. The Binard intrusion continues round the cliff line but at Portaneevey is seen to form only a thin veneer on the vent agglomerate. In the crags north of the observation point the upper part of the intrusion is rubbly, locally reddened and vesicular, and it seems possible that it is in fact a lava flow which forms the dip-slope south of the crest and may be correlated with the lava seen above the Chalk and below the tuff bed east of Portaneevey. Close examination of the cliff, however, fails to show any clear distinction between the supposed lava and the main mass of the intrusion and no flow base can be discerned. It appears probable that the rubbly nature of the upper part of the mass is due to assimilation of xenoliths of agglomerate and that the dip slope is formed by a sill-like part of the intrusion—as is that seen across the laneway to the west.

There are a number of smaller intrusive dolerite masses, including a dyke near the rope bridge and another in the cliff above Casnageeragh.

Not all the intrusions are of the same generation. Tomkeieff and Patterson (1948, p. 208) showed that the Carrickarade intrusion is of Staffa-type basalt while Binard is an olivine-basalt.

Marginal relationships of the vent

The junction of the vent with the Cretaceous country-rocks occurs at Portaneevey on the east and Stackahorlin on the west, but is not exposed at the former locality. Above Stackahorlin the junction of the Chalk and agglomerate rises steeply in the lower part of the cliff and then curves over to the west where the agglomerate overlies the chalk for a short distance. It appears to terminate at a small fault. The westernmost section of dark agglomerate seen is 6 m thick and is overlain by 3 m of drift.

At Portaneevey the contact between the vent and the Chalk was presumed by Tomkeieff and Patterson (1948, p. 205) to cut in turn the Chalk (30 m seen), a single lava flow (9 m) and the chalky tuff (25 m).

In fact there is no exposure of the vent margin, the dolerite and tuff-agglomerate being separated from the Chalk on the foreshore by 10 m or so of shingle, and in the steep slope behind the beach the solid rock south-east of the margin is completely concealed by scree. There is no evidence that the vent 'traverses' the bedded chalky tuff, which can be shown to overlie it.

The reason for the embayment in the cliffs at Portaneevey is not clear, but there is no evidence of a fault at the south-east side of the bay and the Chalk on the foreshore and the base of the lava above the chalky tuff both maintain their level across the scree-masked slope. There is no reason to assume that the succession outside the vent margin is different from that across the bay. There is a small fault, with a throw of a few metres to the south, in the basalt lavas at the north-west end of the cliff which raises the level of the base of the chalky bedded tuff by almost 5 m to the north, but whether this has any connection with the vent margin is not clear.

There is some evidence of a fault in the area of the vent, between Larry Bane Bay and Portaneevey, in that the top of the Chalk in the former area is over 50 m OD while in the latter it is at 25 m. The fault does not affect the bedded tuff and must have preceded its deposition.

MINOR VENTS

On the foreshore just north of The Berrins an elliptical area about 45 m by 29 m shows a jumble of blocks and cobbles of basalt, up to 4 m across, embedded in a soft matrix of comminuted laterite and lithomarge. Some of the blocks show flow-banding on weathered surfaces and are probably derived from the massive flow-banded lava seen on the seaward side of the vent. Within the vent area the flow-banded blocks are random in orientation. The remaining blocks of basalt in the agglomerate are of compact and vesicular basalt. The margins of the vent are not sharp but the agglomerate passes into undisturbed lava over a distance of several metres. Some 18 m south of the main exposure there is a small area athwart the jetty, about 14 m in diameter, showing a similar agglomerate, and 90 m east of the main vent the cliff behind Low Rock Castle shows an apparently vertical vent 15 m wide with irregular masses of basalt in a laterite/lithomarge matrix. Exposures are rather poor but, like the foreshore occurrences, this vent has no sharp margins but passes gradually into undisturbed lava.

140 m south-east of Ringree Point the north face of a raised-beach stack and an extensive area of the foreshore north of it show an agglomerate of varied basalt blocks in a slaggy lateritic matrix [C 833 395]. This area, 64 m by 37 m, is probably a vent.

90 m east of the east side of Port Gallen an elongate area 55 m by 9 m to 14 m running north-west to south-east across the foreshore [C 836 396] has blocks of basalt and of pale-brown recrystallised limestone in a matrix of comminuted material including streaks of laterite. This outcrop has the same direction as the strike of the local lava flows and might be an unusual exposure of the weathered top of a flow as these have, in this locality, occasional blocks of limestone, presumably due to pyroclastic activity. It seems, however, to have a vertical contact with the lava, at least on the south-west side, and is probably a vent.

Where the boundary between the townlands of Craigtown Beg and Craigtown More crosses the foreshore it corresponds almost exactly with the line of a thin dyke. Two irregular vents are seen to west and east of this dyke [C 838 397]. The western, some 45 m in maximum diameter, contains blocks of basalt, some of enormous size, and of pale brown recrystallised limestone, vesicular lithomarge, and red ferruginous laterite, in a matrix of very lateritic composition. The eastern vent, seen only in an islet exposed at low tide, contains a jumble of basalt blocks, from pea-size up to 3 m boulders, in a red lateritic matrix. Both these vents are remarkable for the very abundant laterite, not only in the matrix, but also as blocks in the agglomerates.

Two low-tide islands north-east of Long Port are composed of rounded and subangular blocks of basalt, vesicular and compact, in a matrix of comminuted basalt with flecks of red laterite [C 843 399]. The extent of this vent is uncertain but it seems to be at least 64 m in diameter.

140 m south-east of the Long Port vent the cliff north of Ardmore Cottage shows a section through a vertical pipe of agglomerate 14 m wide, with basalt blocks up to 4 m across [C 844 398]. The exposure is not closely accessible but there is no sign of alteration in the adjacent lavas except for some shattering, on the west side of the vent, in the uppermost flow seen in the cliff.

On the east side of the Devil's Port West, an excellent section is seen through a vertical pipe of agglomerate, 9 m wide at the base and widening somewhat at the top (Plate 19). The agglomerate consists of basalt blocks in a matrix of comminuted basalt, lithomarge and laterite. Three flows of the adjacent basalt lavas can be seen and the levels of the weathered tops of the flows show a displacement of a few metres down to the north across the vent, suggesting that it follows the line of a small fault. The width of the pipe varies, being constricted where it cuts the harder bands of lava and increasing at the softer lithomarge/laterite bands which were less resistant to lateral stoping. In places blocks of the more massive lava can be seen in the pipe, separated from the mass of the flow by thin tuff veins. These have moved only a few centimetres and were presumably in the course of incorporation in the agglomerate when the activity ceased.

This vent may be independent but it may well be the western end of the much larger diatreme which is seen in the cliffs west of East Devil's Port [C 845 398] extending from a point 55 m north of the road as far as the western cave. Junctions are near vertical where seen in the cliffs but are much less clear than at Devil's Port, the blocky agglomerate passing into rather broken basalt lavas over a distance of a few metres. Near the southern side of the vent the agglomerate is streaked with irregular bands of lateritic material which gives an impression of large-scale onion-shell jointing. The lavas seen north of the vents, round the end of the spur between Devil's Port and Old Castle Port, are very broken and rubbly, as are those seen south of the vent in Old Castle Port.

The complex of agglomerate pipes and sheets at the White Rocks is exposed in the cliffs above the beach for over 320 m. The westernmost exposure [C 885 407] is a large stack, 15 m high, of shattered Chalk and flints with irregular bands and pipes of chalky agglomerate including basalt boulders, generally rounded, up to a metre across. These agglomerates are seen in the south face of the stack and down to beach level on the west face. The cliff behind the stack shows a thickness of 18 m of shattered and recemented Chalk with broken flints scattered throughout. At the cliff top the Chalk is overlain by over 2 m of coarse, rubbly basalt, very weathered, which passes under grass. This is presumably a minor intrusion. At its western side this cliff is intersected by an irregular sheet of Chalk/basalt agglomerate which appears to dip east at about 45° and is 3 m or more in thickness. Where this reaches the cliff top it abuts against a sheet of basalt/flint agglomerate, without Chalk, 4.3 m wide, and another basalt/flint sheet or pipe is seen in the shattered Chalk 7 or 8 m to the west.

On the beach just east of the Chalk/agglomerate stack there are occasional outcrops through the sand, over an area 30 m wide and extending at least 40 m seawards from the cliff foot. These exposures are of black basaltic agglomerate, with blocks of compact and vesicular basalt and coarse-grained dolerite, rounded and subangular, in a matrix which includes much red lateritic material. Just beyond this are three sea-stacks of shattered and recemented Chalk, the easternmost of which has occasional blocks of coarse basalt embedded in the chalky matrix. The cliffs behind these stacks show exposures of another vent of basaltic agglomerate. This vent is some 45 m wide from east to west and, though its souther margin is not seen, it is over 14 m wide from north to south. The western margin is seen in the cliff to be stepped, widening upwards. The eastern is vertical; as in the vent on the beach the agglomerate is composed of rounded and subangular blocks of basalt up to 3 m across, and it also includes a mass of rubbly chalk 4 m long and 1 m wide.

18 m east of this vent a small pipe of basalt/flint agglomerate is seen in the top of a stack of shattered Chalk which includes basalt boulders on its east face. A few metres beyond this there are small exposures of basalt agglomerate on a steep bank 15 m above the beach and of basalt/flint agglomerate in a near-vertical sheet at beach level, both in shattered Chalk.

The bluff of chalk beyond these exposures, which forms a spur west of the Cave is shattered and recemented except at its northern end where normal compact Chalk with flint bands forms the lowest 9 m of the spur. Above this level the Chalk is shattered.

Just east of the Cave a 2 m wide pipe or sheet of basalt/flint agglomerate with a chalky matrix is seen in the shattered Chalk, with an offshoot dying out to the west. In the spur east of the Cave, as in that to the west, shattered Chalk passes northwards into normal bedded limestone.

Stradley Rock [C 887 407], a natural arch in a raised-beach stack, is composed of shattered and recemented Chalk and south of it the north edge of another vent is seen where, over an area 23 m wide by 5 m deep, exposures of dark basaltic agglomerate occur among the blown sand. Eastwards the shattered Chalk in the cliffs becomes more compact and finally resumes its normal bedded appearance at Brock's Cove.

140 m east of Brock's Cove, and some 210 m east of the last vent agglomerate, in a spur of completely normal unbroken Chalk a mass of fresh vesicular basalt 3 m by 1.5 m by 1.8 m is seen embedded in apparently undisturbed Chalk 7.6 m above the beach [C 890 408]. It is surrounded by an envelope of decomposed basalt and Chalk fragments, presumably an intrusion breccia, 15 cm to 20 cm thick.

The Priest's Hole [C 892 408] is the collapsed roof of a cave which can be seen from the foreshore to be cut along a series of small crush-zones in the Chalk. When viewed from the road the upper part of the shaft appears to be in a basaltic agglomerate, but in view of the known collapse history of the hole and the fact that the Chalk–basalt junction is only a metre or so above the present top of the shaft it is unlikely that this feature is a true explosion vent. The same arguments apply to two small exposures on the grass-covered slope due south of Jackstone Cove [C 895 409] where blocks of basalt are seen apparently embedded in Chalk 3 m below the base of the lavas.

A series of exposures at the east side of Portnool and in The Riggin [C 897 410] may be of intrusive agglomerates but once again the close proximity of the base of the lavas makes these occurrences doubtful. Above the cave which runs under the spur at the west end of The Riggin large blocks of basalt in a chalk and flint rubble are seen about 8 m below the base of the lavas. Similar occurrences are seen on the north side of The Riggin above the natural arch. 45 m east of this, on the south side of the pool, a vertical shaft in the Chalk about 8 m wide is filled with basaltic rubble and the overlying lavas are much disturbed. None of these exposures is accessible for close examination. The vent noted by Patterson (1962) at Gulls Point has not been confirmed.

Near Dunluce Castle the area known as The Pound [C 905 413] between the castle rock and the mainland cliff is underlain by agglomerate. This is mainly basaltic, with blocks up to 4 m across, but contains appreciable flint and the ashy matrix includes much comminuted chalk in places. It has been claimed (Patterson, 1962) that this vent extends over the Castle Rock, an area of the mainland almost as far west as Burnfoot, and much of the foreshore at Portnalea, with vent walls delimiting the foreshore areas of Chalk at Portnalea and east of Dunluce Cove. This 'vent' includes an area of basalt lavas regarded as foundered in the caldera, and covers a total area of some 33 000 m² above high-water mark. The Castle Rock appears to be built up from very rubbly lavas, and at the east side, and apparently overlying the Chalk on the foreshore, these are underlain by a bed up to 8 m thick of what may be agglomerate. This bed, which has a reddened top, is composed of rounded and subangular basalt blocks, generally of moderate size, in a dark basaltic matrix. Patterson states that blocks of recrystallised limestone have been seen in this deposit but its pyroclastic origin is not certain and it might be a ball or block lava. What is apparently the same bed is seen on the foreshore west of the Castle Rock in Portnalea. Though the basalts overlying this deposit are much broken and irregularly weathered, it is not considered that either the Castle Rock or the cliffs south of Portnalea are part of the vent, which is thought to be limited to an area of about 90 m by 45 m, between two faults at The Pound. It must be recorded, however, that W. B. Wright in an unpublished revision survey of the area in 1916, described the Dunluce vent in much the same area as Patterson.

Some 400 m north-east of Dunluce the cliffs at Gortnabane Cove [C 908 416] show an excellent section through a vertical vent of basaltic agglomerate 8 m to 9 m in diameter with blocks up to 2 m long, though generally much smaller, in a greenish calcareous matrix. This vent is seen for a vertical distance of some 15 m, punched cleanly through a banded pahoehoe lava and an overlying massive flow.

730 m east-north-east of Gortnabane the stack of Lacknamodeen [C 915 419], 11 m high, and the foreshore for some distance around it are composed of blocks of compact and vesicular basalt in a matrix of comminuted basalt, lithomarge, and red laterite. There is a gradual transition to the normal lavas at the edge of the brecciated zone. The surrounding lavas over an area 180 m across tend to dip towards Lacknamodeen, the dips being up to 30°. This feature, with its extensive aureole of tilted lavas, suggests a major collapse feature but it has been regarded as a vent by Wright, in his 1916 revision, and Patterson (1962).

On the west side of Portcoon, 550 m north-west of Causeway Head, a small pipe of apparently basaltic agglomerate is seen cutting the lavas near the cliff edge [C 939 440]. The inclination of this feature, which is about 4.5 m in diameter, is not determinable. There are a considerable number of dykes in this area and it seems possible that this is a small rootless vent which resulted from phreatic activity.

VOLCANIC PLUGS

The Bendoo plug at Ballintoy Harbour is a vertical near-cylindrical mass which is well displayed only in the cliff on its northern side. Exposure at beach level shows the western margin to be intrusive but with very little alteration of the adjacent Chalk. The eastern margin is faulted. There are exposures at a higher level beside the road from the harbour and around the Coastguard Cottages which allow the rock in the centre of the plug to be examined.

The intrusion varies from a coarse dolerite in the centre to a fine-grained basalt at the margins. Local bands of amygdaloidal material are seen beside the road, some distance in from the margin.

The absence of alteration in the Chalk beside the intrusion and the radiating joints, which suggest steady cooling, indicate that either the plug was formed by one brief period of activity or that it is a 'blind' intrusion which did not reach the surface and never acted as a conduit. A further possibility is that it, like some other Antrim plugs, may have been the solidified lava-filling of a pit crater—features which are common on the Hawaiian shield volcanoes.

SILLS

Portrush Sill

What may be the southernmost extension of this intrusion outcropped in the roadway of Hopefield Avenue 400 m south of Crock-na-mac Road, where a rounded surface of coarse dolerite was seen in 1957. Only a few centimetres was exposed but the rock is identical with that of the sill.

Massive dolerite underlies the roadway in Causeway Street and is exposed at one point below the front of a house on the west side of the road. From this point northwards the whole town rests on solid rock with only a veneer of sandy soil.

Rock is first seen on the foreshore about 140 m south of Rock Ryan where vesicular coarse-grained dolerite with undulating flat-lying joints is seen on the beach and in the foundation of the sea wall. At the Arcadia Ballroom fine-grained dolerite with small xenoliths of hornfelsed shale is seen below the foundations and at Rock Ryan lenticular slabs of hornfels up to 1 m thick are interbedded with the dolerite. The rocks on the foreshore north of the small sandy beach are of coarse ophitic dolerite and just south of the boat-slip at Portneen a 3.6-m face shows that the massive rock is intersected by bands and veins of more resistant fine-grained material which stands out on weathered surfaces. Wisps and lenses of hornfelsed shale occur in the topmost part of this section and some flow-banding is seen on the weathered face. At the boat-slip conspicuous vertical and horizontal veining is seen in the massive dolerite.

Between Portneen and the Blue Pool there are extensive outdrops of massive dolerite, sometimes with close vertical jointing but generally with undulating gently inclined joints. About 165 m north of Portneen an extensive area is covered by hornfelsed shales. Up to 2.5 m of hornfels is seen to rest on the very irregular surface of the dolerite.

Apart from the area immediately below this hornfels outlier, where the dolerite is chilled and fine-grained, the rock is coarse-grained though it contains blocks and wisps of hornfels and must be near the roof of the sill.

In the immediate vicinity of the Blue Pool a mass of hornfels rests on the very irregular surface of the sill, the top of which is seen just above high water in the small bay beside the pump-house. The greatest thickness of hornfels seen is about 8 m capped by a 0.3-m sheet of fine dolerite. Another dolerite sheet, thinning from 1.2 m to 0.3 m and dipping east at 45°, is seen traversing the hornfels from west to east.

From the Blue Pool to Portandoo Harbour the irregular chilled surface of the sill is seen dipping east on the foreshore and is overlain by extensive slabs of hornfels. At the south end of the harbour thin sheets of fine dolerite are interbedded with the hornfels. Just north of the breakwater a slab of hornfels 2 m thick overlies the chilled top of the sill and is intersected by an irregular basalt dyke 0.3 m to 1 m wide. This cannot be followed into the sill and may be a vertical offshoot from it.

From the breakwater to Reviggerly extensive outcrops of the upper part of the sill, with extensive veining and some hornfels rafts, form a ridge which falls away to Portscaddan on the west. The upper surface of the intrusion is very irregular here for at Reviggerly Point hornfels overlies the dolerite with a near-horizontal contact but 23 m to the south the junction suddenly plunges to the south-west at 70° to 90° and this vertical and overturned contact, with the hornfels forming a wall along the eastern side of the promontory, can be followed for about 55 m before passing out to sea. While the general line of the junction is clear, detailed examination shows small-scale irregularities including small xenoliths of hornfels in the dolerite. Nearly all sedimentary structure is lost but small areas sometimes reveal intensely contorted bedding. The dolerite at the contact is chilled and so fine in grain that it looks very similar to the hornfels, giving the impression that the actual junction is a few centimetres further west than is actually the case. All along this contact the dolerite is very strongly jointed with close vertical joints, normal to the contact, persisting for up to 18 m from it. The intensity of jointing is such that close to the contact it looks like foliation in a metamorphosed rock.

The dolerite forming the upper part of the ridge is massive and apparently homogeneous for 6 m or so, with a variety of coarse and fine veins penetrating it. Below this upper zone the west face of the promontory shows irregular flat jointing dipping east at 10° to 15° with the dolerite in the form of lenticular masses. These are all much weathered and differ in texture and appearance, including a variety with dark out-weathering spots a centimetre or so in diameter in a paler matrix. Progressively lower layers in the intrusion are laid bare on the foreshore west towards Portscaddan, where massive, very coarse, dolerite has occasional lighter, more feldspathic patches with dark-weathering mafic clots. Below high-water mark this rock has a curious rubbly-weathering surface, occasionally cavernous. Jointing is, as in the overlying beds, mainly undulose and horizontal or gently dipping. Vertical joints are random in direction.

In Portscaddan a crush zone 9 m wide probably affects the succession but between here and the Wash Tub, the line of a 2-m crush zone, massive dark coarse dolerite with rare dark mafic clots and rudely columnar jointing is seen on the foreshore.

Ramore Head, west of the Wash Tub crush zone, rises to about 28 m above low-water and is flanked by a wave-cut rock platform which allows access to the lower part of the cliff. The cliffs consist of rude columns plunging west-south-west at 80°. All the rock above water is much weathered by spray, so that its lower part is covered with pea-sized knobs of out-weathered augite. Some patches and bands are particularly liable to weathering to a greenish sand with conspicuous green pseudomorphs after olivine, notably a 3-m band just above the rock platform which can be traced for over 40 m. Horizontal etching on weathered faces is probably due to flow-banding as it occasionally shows overturning of the banding. In places the surface of the rock platform shows projections, often in the form of concentric circles, of more resistant unweathered rock. This material is a feldspar-rich segregation with abundant andesine and is discussed in the section on petrography (p. 140).

To the east of the Harbour extensive quarries, developed during the construction of the breakwaters, are now partially occupied by houses. Faces of up to 9 m of massive dolerite are capped by hornfels, the contact being rather undulating, and there is some close vertical jointing in the upper part of the dolerite. The top 2 m or so is chilled and weathers differently from the rest of the sill. Veins and horizontal sheets of feldspar-augite pegmatite are common in the upper part of the dolerite. The occurrence of the top of the intrusion at about 15 m OD in these quarries suggests that the dip of the sill in this area is about 5° to the east.

The general succession in the mainland part of the intrusion comprises 6 m to 9 m of massive dolerite chilled at the top, over some 12 m of coarse dolerite with horizontal lenticular jointing and mafic clots. As none of this is recognised in the Ramore Head cliffs it is clear that the crush-zones at Portscaddan and the Wash Tub must run across to the Dock, accounting for the hollow of the Recreation Ground, and have a downthrow to the east. The more compact, poorly columnar mass of Ramore Head, 28 m or more thick, is thus the lowest part of the intrusion seen and the total thickness exposed on the Portrush promontory is at least 45 m, with the base unseen.

The group of islands, including North and South Island Ean, Middle Isle, West Isle, Big Carr and Little Carr and the Otter Isles, show disconnected portions of the same general succession. Hornfels overlying chilled dolerite is seen on South Island Ean, and here the upper massive dolerite is about 3 m thick and overlies more than 3 m of rubbly-weathering dolerite with mafic clots. The upper part of this band shows excellent flow-banding on sea-etched surfaces which dip south-east at 10°. All the other islands in this group show spotted dolerite with mafic clots except the northernmost of the Otter Isles where the more massive bottom dolerite appears.

On Little Skerries thin outliers of hornfels are seen on the south-east coast. Below this an excellent section through the sill, now dipping south-south-east, can be seen extending across the island. Massive homogeneous dolerite, heavily veined in places, and sometimes showing flaggy jointing, varies in thickness from 5 m to 8 m. The top metre is chilled and fine in grain and the lowest 0.3 m to 0.6 m is very rough-weathering with knobs of augite. Below this is irregularly lenticular-jointed feldspathic rock varying enormously in granularity over as little as a few centimetres though there is no sign of internal contacts, all the varieties grading into each other. Dark mafic clots are common but patchily distributed, while the groundmass is often noticeably zeolitic in its coarser grades. This band is about 6 m thick and passes down into a zone of massive dolerite full of mafic clots which forms a line of crags across the island. This rock weathers to a slightly carious surface but stands out in compact bluffs about 3 m high. The lowest 0.6 m shows fairly regular layers a few centimetres thick, due to the presence of horizontal sheets of brown-weathering material. Similar brown-weathering material forms occasional veins 3 cm to 5 cm thick, seen in the massive dolerite crags.

Below the massive dolerite band there is about 6 m of carious-weathering rubbly spotted dolerite with mafic clots generally 5 cm to 8 cm in diameter. This zone also has irregular flat-lying joints but unlike all the overlying zones has no veining by late-stage segregation products.

The lowest part of the sill exposed here, forming the north-west coast of the island, is massive slabby-jointed dolerite, with a few mafic clots which do not stand out on weathered surfaces as they do in higher zones. There are occasional carious-weathering patches up to 0.5 m or so in diameter but they are apparently identical with the rest of the rock save that pyroxene occurs in smaller grains. About 8 m of this zone is exposed.

The total thickness of the sill exposed in Little Skerries is thus about 30 m.

On Winkle Island, only some 200 m east of Little Skerries, the proportions of the various zones of the sill have changed considerably. Two outliers of hornfels are seen on the south shore. The western, over 6 m thick, has a vertical contact with the dolerite on its north-east side but overlies it normally on the east. A smaller outlier 2.5 m thick, about 18 m east of this, is about 5 m higher. The top of the sill here is obviously irregular. The compact massive upper zone of the intrusion, traversed by veins and sheets of segregative and mobilised material, forms the higher ground at the south end of the Island dipping south at 10°. It is underlain by very coarse dolerite with mafic clots at top and base and irregular pillow-jointing. There are some sheets of brown-weathering material 3 cm to 5 cm thick, and the whole zone is some 4 m thick. Below this, pale-weathering, coarse dolerite with irregular mafic segregations and irregular horizontal jointing underlies the 'waist' of the island which is covered by the sea at high tide. The northern part of the island is of massive compact and fairly homogeneous dolerite with small dark mafic clots, but shows no other signs of segregation, and no veining. If the 10° dip persists over this part of the island the thickness of this zone must be over 20 m.

Castle Island is so called because of the occurrence of a large stack of hornfels near the north shore. The eastern half of the island is underlain by massive hornfels dipping south at 10° to 20°. The hornfels–dolerite contact along the north coast is vertical or near vertical with vertical rafts of hornfels in the sill parallel to the junction. The sill is strongly jointed with close parallel joints normal to the contact. The 'Castle', and the ridge running south-east from it, consist of vertical and inverted hornfelsed shales at a lower level than the base of the main hornfels outlier. They must represent a great slab of shale torn off and enveloped in the dolerite which is all chilled and fine-grained in this area. The western promontory is all in the massive upper layer of the intrusion, with many veins of coarse and fine material. An outlying patch of hornfels is seen near the western end.

At the eastern end of Castle Island, and on the adjoining islet, a large vein of fine-grained material with coarser margins on both sides, or sometimes only on one side, cuts the dolerite with a north-west trend. This vein is 0.6 m to 0.9 m wide and the central portion is apparently a tuffsite, similar to intrusive tuffs described from Rhum in the Inner Hebrides (Hughes, 1960).

Large Skerries, the largest island of the group, has a steep north-facing cliff and a dip slope, partly covered by vegetation, which corresponds approximately to the top of the sill. Fairly extensive outliers of hornfels on the south coast near the centre of the island reach a thickness of 6 m. On the western side of the western outlier the dolerite/hornfels contact plunges nearly vertically, and on the east side of the eastern outlier the contact plunges steeply to the south-west. In the rest of the outliers the junction seems to be more regular and dips south at about 12°. All along the south coast of the island, and in exposures on the dip-slope, the dolerite shows the typical upper-surface features seen elsewhere. Small outlying patches of hornfels a few centimetres thick, lenses of chilled tachylitic basalt, extensive late-stage veining with veins and sheets of coarse and fine-grained material, and cuboidal small-scale jointing, mirroring that which occurs in the hornfels, are all common. Near the north-east corner of the island an exceptional sheet of late-stage segregation material is intruded 2.4 m below the top of the dolerite and reaches a thickness of 1 m. This sheet itself shows a 3-layer segregation into an ophitic pyroxenite, a central zeolite-gabbro, and a basal pyroxene-rich gabbro.

Also in the north-east corner, deeply under-cut joints, one of which forms a feature running from north-east to south-west across the island, allow the sea to penetrate far inland.

The cliffs on the north side of the island which rise to about 27 m OD display the same general section through the upper part of the sill as has been described from the other islands. At the west end about 6 m of pale-weathering massive dolerite overlies some 12 m of rough-weathering spotted rock. The upper half of this zone is marked by irregular flat joints giving the effect of a pile of flattened pillows in the cliff face. The dark mafic clots are most abundant in the lower part of the zone, sometimes in aggregates, often scattered. Generally about 5 cm in diameter, they are sometimes up to 15 cm or down to 1 cm. Occasional coalescent groups of large irregular clots give a very rough surface on weathered faces. At the base of the cliff, 6 m or so of compact dolerite is seen above sea level.

In the centre of the island the cliffs are largely inaccessible but a well-marked layered appearance is common, with a massive top layer over a rough-weathering dark zone.

The best sections are those at the east end where 6 m of massive, homogeneous, rather pale dolerite passes down into 1.5 m of similar massive rock with abundant mafic clots about 3 cm in diameter. This passes sharply into 9 m of carious-weathering dolerite with mafic clots and bands of darker and more mafic material. The lowest 3 m or so of the cliff is of more massive and homogeneous material.

Black Rock consists of two islets, the easternmost of the Skerries group. The southern, of massive dolerite, is part of the top layer of the sill, while the northern is rubbly-weathered and is part of the central zone.

Knocksoghey Sill

The intrusion is best exposed in the old quarry on the north side of the hill, once worked as a source of square setts for street paving. About 20 m of rudely columnar dolerite is seen in the face, but the total thickness of the sill is about 38 m.

The base of the intrusion, which was injected between the overlying Lower Basalts and the underlying tuff cone, is seen at a few points immediately west of the quarry and is apparently horizontal. The dolerite is fine-grained and amygdaloidal for about a metre from the contact, the vesicles being stretched out parallel to the contact and filled with zeolites. The underlying tuff has been contact-altered for a few centimetres with close-spaced joint planes. The top contact with the lavas is not exposed.

The sill has conspicuous phenocrysts of yellow olivine at some levels but there is no sign of gravitational layering. There are bands of zeolite-rich rock which are narrow and have sharp but gradational contacts with the surrounding dolerite. Irregular sheets of dolerite pegmatite intersect the dolerite locally, and there are also thin veins of reddened dolerite of uncertain origin.

Evidence for the intrusive nature of this feature is its massive character and irregular jointing, the apparently intrusive nature of the basal contact, and the petrographic similarity of the rock to material from other sills in the area. It certainly has all the appearance of a sill. The occurrence of basalt pegmatite, adduced as an additional positive factor by Tomkeieff and Patterson (1947), is not conclusive, as such pegmatites are now known to occur in massive lava flows in this district.

DYKES

On the foreshore between the townlands of Craigtown More and Craigtown Beg [C 837 397] a dyke of fine-grained olivine-basalt from 0.2 m to 0.8 m in width cuts the basalt lavas between two volcanic vents. At its seaward end the dyke has a sinuous course and sends small tongues of basalt into the adjacent lava at several points.

Just north of the Breakwater at Portandoo Harbour [C 856 413] a dyke of fine-grained basalt with an irregular course forms an L-shaped outcrop in the hornfels. It varies from 0.3 m to 0.9 m in width and has horizontal jointing. About 200 m to the west a hornfels exposure at the north end of Lansdowne Crescent shows a vertical contact against decomposed dolerite of which 0.66 m is exposed. This may be a dyke though it is probably part of the sill.

At the west end of Large Skerries [C 872 426] an irregular basalt dyke runs north-west across the island. It incorporates blocks and sub-parallel sheets of very fine-grained basalt which may represent spalled-off slivers of the chilled top of the sill. Varying from 0.6 m to 0.3 m in width, the dyke can be followed to the top of the cliff at the north of the island where its course becomes very irregular and it thins at places to 10 cm. It is not seen on the rock platform at the base of the cliff. An irregular north-east-trending dyke about 0.3 m wide runs across Castle Island, intersecting both sill and hornfels, and a small 0.2 m dyke trending north-west cuts the hornfels on Winkle Island.

At Brock's Cave, White Rocks [C 887 407], a cave in the Chalk 73 m long has been cut along an irregular dyke, 0.6–0.9 m wide, of completely decomposed fine-grained vesicular basalt.

The cliff and foreshore east of Portcregcarragh [C 918 421] show the lavas cut by a vertical but irregular dyke to 1.2 m wide of porphyritic basalt. 275 m north-east of this, an inclined sheet dipping south-west at 40° is seen on the foreshore. This dyke is 1.2 m to 1.8 m thick and has chilled selvages.

A thin (0.3 m to 0.6 m) dyke-like sheet, locally flattening into a sill, is seen at Seagull Rock, Portballintrae, running in a north-east direction and an irregular 0.3 m to 2 m dyke is exposed on the foreshore at Salmon Rock.

At Bush Foot a narrow dyke 0.5 m wide crops out on the west side of the river while a massive intrusion at least 9 m wide is seen as a stack on Bushfoot strand.

North of Runkerry House two parallel dykes about 3 m wide are seen at Polbrinck and another pair 1.5 m wide are exposed on the foreshore 130 m to the north.

The tidal rock called The Milestone [C 935 439] is part of a massive dyke seen on the foreshore at Runkerry Point where it is 5 m wide.

A 1.8-m north-westerly dyke with horizontal columnar jointing is seen just west of Leckilroy Cove. On the wide raised-beach rock platform east of the Cove two parallel dykes trending north-north-west redden the country rocks. The western is irregular and about 0.6 m wide: the eastern, only a few metres away, is about 1.8 m wide. Seagull Isle [C 939 442] at the entrance to Portcoon is a stack formed by two parallel massive horizontal-jointed dykes 3 m and 2.4 m wide and 0.6 m apart. On the west side of Portcoon an irregular slaggy looking dyke is seen in the cliff, 1.8 m wide at the bottom and ramifying into thin impersistent veins upwards.

90 m east of Portcoon Cave [C 944 446] a very irregular ramifying dyke of fine-grained tholeiitic basalt is seen on the top of the raised-beach rock platform. Varying in width from a half a metre to two metres, this intrusion encloses, or envelops, large masses of rather brecciated basalt. On the southern side of Portnaboe the 'Camel Rock' [C 945 446] (not named on the map) is a raised-beach stack about 12 m high formed by the weathering out of a massive dyke-form intrusion. For some 40 m at its northern or seaward end this intrusion is seen on the rock platform to be an irregular squirt about 0.3 m wide with incorporated blocks of lava and minor ramifications. It widens gradually southwards and suddenly develops good horizontal jointing which is sometimes curved and often spectacular. The dyke reaches a maximum thickness of 6 m with vertical zones of heavy calcite veining. There is some reddening of the adjacent country rock. At the landward end of the stack, near the Slip, the dyke crosses the line of a small north–south fault and side-steps or is displaced some 3 m to the south. Its extension southward is marked by two exposures of horizontally-jointed dolerite on the grassy slopes below the Causeway Hotel. About 90 m east of the Slip an irregular dyke only about 10 cm wide is seen on the foreshore for about 7 m.

The three divisions of the Giant's Causeway are separated by two north-westerly dykes of fine-grained dolerite. The western, between the Little and Middle Causeways, is poorly exposed at low water where it is seen to have horizontal and longitudinal joints. The eastern dyke, 1.8 m wide, is seen at high-water mark and is also exposed in the bank above The Loom. It has well-developed horizontal columnar joints. Neither dyke is seen in the cliff behind the Causeway.

Three north-westerly dykes are seen on the foreshore at Port Noffer, and, beyond the obscuring scree slopes, all are exposed in the cliffs. The westernmost outcrops on the foreshore about 180 m east of The Loom and is some 5 m wide with a slightly irregular course. It is seen in the cliff 65 m east of the Shepherd Path gate. The central zone of this dyke is closely cross-jointed but marginal zones 0.3 m to 0.6 m wide are longitudinally jointed and show vertical banding of vesicular and compact material. The boundary of these marginal zones is often very irregular. 290 m east of The Loom a very irregular 1.8 m dyke is seen on the foreshore. This dyke is of fine-grained feldsparphyric basalt and is cross-jointed with no marginal variation: at the seaward end of the exposure it is seen to incorporate a block of basalt lava. It is poorly exposed in the cliff. A 3-m dyke of cross-jointed medium-grained basalt is seen on the foreshore below The Organ [C 951 448] and is well exposed in the cliff south of The Organ.

At Roveran Valley Head [C 952 452] a massive north-westerly dyke of fine-grained tholeiitic basalt 5 m wide forms an offshore islet and is seen in the cliff below and above the path. Where it intersects the Interbasaltic Bed the dyke, and the adjacent country rock, for up to 15 m to the east, are intersected by rami-fying squirts of basalt up to 0.3 m wide, parallel to the dyke. Above the path the dyke transgresses eastward and forms the west face of the headland where its rather irregular course and random jointing can easily be seen.

In Port-na-Spaniagh three dykes are seen in the cliff. The westernmost is about 3 m wide but inaccessible and is not seen on the foreshore. The others trend north-north-west and are seen on the foreshore to converge, but not to coalesce, at low water. The western is 3 m wide with strong longitudinal jointing while the eastern is 6 m wide with strong horizontal honeycomb jointing and incorporates blocks of country rock. Both dykes are of olivine-basalt.

At Port na Tober [C 962 456], just west of Pleaskin Head, a 3-m dyke of horizontal-jointed olivine-basalt is seen on the foreshore and in the cliff, trending north-north-west.

At White Park Bay ten dykes ranging from 2 m to 4 m in width are seen in the White Limestone cliffs at the back of the landslip, six of them in the 400-m stretch west of Ballintoy School. Five dykes which appear on the beach intersecting Lias mudstone may be continuations of the cliff exposures though they are narrower —1 m to 2 m. The largest beach exposure is the Long Causeway, a 2-m dyke, bifurcating to the south, of coarse dolerite, amygdaloidal in the centre.

Between White Park Bay and Boheeshane Bay there are a further six dykes. A 1.2-m dyke is seen in the cliff south of Dunshammer, but not on the foreshore, and a 2-m dyke trending west-north-west intersects pyroclastic rocks at Stackandoo. Another 2-m dyke is seen in the old chalk quarry south of Portnalug, and a narrower (0.6 m) but bifurcating dyke is seen cutting the chalk 100 m north of the quarry exposure. The small chalk quarry below the road and behind the limekiln at the harbour has one or possibly two irregular dykes, now poorly exposed.

The Bendoo plug has two dyke-like intrusions which may, perhaps, be portions of a single curved intrusion. A 4-m dyke is seen in the north face of the plug and a 5-m dyke in the east face.

Three dykes intersect the chalk in the cliffs behind Larry Bane Bay, the largest some 4 m wide, and a dyke is also seen intersecting the tuff bed 250 m south-south-west of Stackahorlin.

Three dykes are recorded in the now-overgrown Chalk quarry [D 071 349] in the south-east corner of the sheet. Still visible are a bifurcating basalt dyke 5 m wide and trending north-north-west seen below the ruined house in the south-west corner of the workings, and a small exposure of coarse basalt seen below Chalk in the north-east face which may be connected with a north–south dyke shown in this position on the old field map. No other signs of this dyke, or of a converging north-west-trending dyke also recorded by the surveyor of 1885, are now to be seen.

PETROLOGY OF THE INTRUSIVE IGNEOUS ROCKS

Vents

The tuffaceous matrix of the vent fillings is commonly decomposed and difficult to examine, and few thin sections have been made from it. In the smaller vents it appears to be largely comminuted decomposed basalt and laterite. The material from the Carrickarade vent, however, consists of scattered serpentinised olivine crystals up to 0.8 mm in diameter, set in a matrix of partially welded irregular fragments of basaltic lava. The lava is composed of finely divided feldspar, ferromagnesian grains, granular opaque matter and some glassy material. A few porphyritic laths of feldspar occur and patches of zeolite occur between many of the lava fragments (NI 1534). This tuff is notably different from that which occurs in the bedded pyroclastics which surround the vent (p. 131).

The vent intrusions at Carrickarade are of two types. That on the mainland—the Binard Intrusion of Tomkeieff and Patterson (1948)—is an olivine-basalt, varying considerably in grain size from place to place, while the main intrusion on the island is low in olivine and approaches tholeiitic basalt in composition.

The mainland intrusions are of variable olivine-basalt. One specimen (NI 1517) consists of ophitic plates of augite enclosing labradorite laths, the feldspar showing local alteration to chlorite. Phenocrysts of olivine are all very altered and the mesostasis consists of fibrous zeolite and chlorite, the latter enclosing glass fragments. A second specimen is of non-ophitic texture, with scattered phenocrysts of serpentinised olivine and feldspar laths set in a granular matrix of plagioclase, augite, serpentinised olivine and magnetite. Thomsonite occurs in amygdales and veins (NI 1536). Tomkeieff and Patterson (1948, p. 209) describe a taxitic phase from this intrusion, with patches in the rock consisting of fresh euhedral grains of pyroxene, labradorite and iron ore set in a groundmass of zeolites—chabazite, thomsonite, gmelinite and natrolite—which appear to be magmatic in origin and not secondary vesicular minerals.

The intrusions on Carrickarade island are poor in olivine—Tomkeieff and Patterson estimate 4 per cent; normal olivine-basalts in Antrim average 15 per cent—and the sparse olivine grains, partially altered or fresh, are set in a matrix of small labradorite laths and granular augite and magnetite. Some glassy mesostasis is present (NI 1533, 1535). This rock approaches the composition of the Tholeiitic Basalts, but in its olivine content bears an affinity to the Staffa basalts of the Hebrides.

Plugs

The rock in the Bendoo Plug varies from a coarse dolerite in the centre to a fine-grained marginal facies. The dolerite has fresh olivine, ophitic towards some of the plagioclase, and large plates of fresh ophitic titanaugite. There is a zeolitic mesostasis with some chlorite.

The small plug at Ballymacrea is an ophimottled olivine-basalt with pale green olivine, ophitic augite, labradorite laths and magnetite. It is, like the Bendoo plug, not petrologically distinctive (NI 1393).

Sills

The petrology of the Portrush Sill is discussed in detail in a separate paper (Hawkes and Wilson, 1975) and will therefore only be summarised here. Similarly the sills penetrated by the Port More Borehole will be described in detail elsewhere.

Portrush Sill

Only the upper part of this sill is exposed and consists of a generally massive and veined top zone with chilled top; a central zone characterised by segregative clots of olivine, sometimes with a middle band of more massive material; and a lower zone o massive dolerite with very rare segregationary features.

The top zone, generally about 6 m thick, is of sub-ophitic or ophitic dolerite with a chilled top against the overlying hornfels. The contact between hornfels and basalt selvage is marked by a narrow zone of a colourless prismatic mineral of high refractive index (NI 1382). The chilled basalt is very fine-grained and consists of small microlites of untwinned plagioclase, minute phenocrysts of labradorite, granular magnetite, abundant granular pyroxene, and occasional small phenocrysts of corroded olivine. The feldspar phenocrysts are often in stellate or cruciform aggregates (NI 1382). In another locality interstitial analcite is present in a rather coarser-grained basalt (NI 1452). A specimen from Castle Island (NI 1434) shows the pyroxene (augite) to increase towards the contact until at the junction the rock is approximately 90 per cent granular augite with a little intergranular feldspar.

The topmost metre or so of the dolerite is appreciably finer in grain than the rest of the zone (average grain-size about 6 mm). It is a sub-ophitic rock with twinned labradorite in long acicular grains, titaniferous augite and partially serpentinised olivine. Both augite and olivine are ophitic to the feldspar (NI 1376, 1423). The main mass of the upper zone is a coarse ophitic dolerite with augite plates up to 3 mm or more in diameter. Olivine appears to be patchy in distribution: some slides show little of it while in others it is abundant and ophitic to feldspar. Magnetite in large grains is a common accessory (NI 1424a, 1441). An unusual type from this zone is a specimen from Reviggerly (NI 1383) in which the olivines are partially resorbed and the feldspar laths are partially replaced by analcite. A little acid plagioclase occurs interstitially. This rock appears to approach the type described as 'poikilo-plektophitic' by Harris (1937, p. 104).

There is a gradual change in the character of the top zone over a distance of half a metre or so into a paler and rough-weathering central zone characterised by the presence of dark clots which stand out on weathered surfaces. The groundmass of this spotted rock is variable but is often marked by the occurrence of quartz and a granophyric texture. A specimen from the mainland at Portscaddan is a highly altered quartz-dolerite with subhedral oligoclase in an ophitic groundmass of pale green diopside. There is some interstitial analcite and quartz and analcite also occurs abundantly as a replacement for the feldspar (NI 1371). On Large Skerries the rock some 3 m below the top zone is a feldspathic dolerite consisting dominantly of subhedral oligoclase channelled by analcite (NI 1415). A specimen from only 1.5 m higher is of a normal ophitic olivine-dolerite with labradorite and augite, but this band, though marked by dark mafic segregations, is similar to the top zone in weathering characteristics and is an intermediate type. Lower in the middle zone a specimen from the rock platform on Large Skerries (NI 1426) is of slightly granophyric dolerite, consisting of labradorite zoned to oligoclase, and augite with magnetite and analcite. The pyroxene grains commonly have an outer zone of aegirine-augite not quite in optical continuity with the core, locally partly replaced by biotite. Granophyric outgrowths surround many of the feldspar grains and biotite and iron oxides are also associated with these intergrowths.

On Little Skerries the upper part of the central zone is very variable in grain size and a band of more massive material forms a feature across the island. The variable material is a feldspathic granophyric dolerite, the coarse variety (NI 1446) differing from the finer only in lack of secondary analcite which in the latter (NI 1447) channels the labradorite. The massive band (NI 1448) is a similar granophyric dolerite extensively analcitised. Much of the feldspar is in the form of granophyric intergrowths of feldspar and quartz.

All the specimens examined from the central zone are wholly deficient in olivine, which occurs only in the form of glomerophyric aggregates. These dark-weathering 'clots' consist of olivine generally more or less altered to brownish-red 'iddingsite' or goethite, and labradorite, with accessory amounts of augite and magnetite (NI 1416, 1427). The labradorite is sometimes enclosed by the olivine and occasionally veined by its alteration products (NI 1445). A specimen (NI 1371A) from Portscaddan has the appearance of a normal olivine-dolerite without any preponderance of olivine. The feldspar is labradorite, unlike that of the surrounding country-rock which is oligoclase (NI 1371).

The lowest zone seen of the sill is massive and fairly homogeneous dolerite. That from the mainland (at Ramore Head) is an ophitic olivine-dolerite (NI 1373) with occasionally harder segregations in the form of concentric sheets which are more feldspathic than the country rock and contain basic andesine as their dominant feldspar instead of the normal labradorite (NI 1372).

On the Skerries, however, the lowest zone is remarkable for its notably feldspathic dolerite, with scarce olivine and augite phenocrysts in a labradorite-rock (NI 1437). On sea-weathered surfaces this rock sometimes shows cavernous-weathering patches which appear to be identical to the rest of the rock save for the occurrence of the augite in smaller grains (NI 1443).

The only specimen available from the group of intrusions interdigitated with the Chalk in the borehole at Corbally reservoir is an olivine-free dolerite with phenocrysts of andesine, acicular labradorite, partially chloritised, and sub-ophitic augite. Though this is not closely comparable with any material examined from the Portrush Sill there is so much variation in the main sill that it is possible that the Corbally intrusions are indeed an extension of the intrusion to the south, perhaps with some contamination by the limestone.

VEINS The very abundant late-stage veining of the sill was studied in detail by Harris (1937) who divided the veins into six main groups:

A   Veins of mobilised hornfels
B   Plagioclase–pyroxene veins and sheets with or without olivine
   1 Intergranular basalt
   2 Orthophyric veins
   3 Hypersthene-dolerite and olivine-dolerite with poikilo-plekophitic texture
   4 Dolerite pegmatite and leucocratic porphyrite
C   Calcite–chlorite–zeolite veins

Harris suggested that all the material of Group B, including the most abundant type—B4—was of syntectic origin, derived from the overlying hornfelsed shale by the agency of highly energised emanations from the intrusion. This thesis may be true for groups B1–B3 but seems unlikely to be of universal application as veins of dolerite pegmatite are common in other major Tertiary intrusions—the Fair Head, Scrabo and Knocksoghey sills, for example. No attempt has been made to recover the localities collected in detail by Harris on the Portrush promontory and only a few sections have been made from material collected on the mainland. An extensive collection was made, however, on the islands of the Skerries.

No veins of mobilised hornfels were found in the collections, but a representative of the Harris group B1 is a sheet of hornfels-like material, apparently a xenolith, within 30 cm or so of the top of the sill on the Large Skerries (NI 1422). This specimen has a fine-grained edge in contact with the dolerite which is a tachylite consisting of skeletal laths of labradorite averaging 0.05 mm in a groundmass of granular augite with some magnetite. The bulk of the sheet is a spherulitic basalt with skeletal microphenocrysts of labradorite in a groundmass of spherulitic fibrous feldspar with interstitial augite granules. There seems no doubt that this sheet is a metasomatised hornfels raft.

Veins of granophyric dolerite, which weather with a characteristic brown crust, have been collected from the top of the sill on the Large Skerries (NI 1410, 1421), from the lower part of the upper zone on Winkle Island (NI 1439) and South Island Ean (NI 1453), and from the central zone on Little Skerries, 15 m below the top of the sill (NI 1449). These veins, which are the equivalent of Harris's Orthophyric Veins (B2), are characterised by the occurrence of tabular oligoclase, often zoned, and free quartz which appears as grains in the matrix, amygdale fillings, and myrmekitic intergrowths with feldspar and other minerals. The feldspar in the groundmass is usually albite and biotite is a common accessory. The pyroxene is sometimes greenish augite (NI 1449) but may be colourless (NI 1410, 1421). Olivine completely pseudomorphed by iddingsite occurs in one specimen (NI 1449) and other accessory minerals recorded are apatite, analcite, fibrous zeolites (mainly natrolite), and iron ores.

Two specimens from Reviggerly (NI 1374–5) one from Castle Island (NI 1432) and six from the east end of Large Skerries (NI 1408, 1411, 1412, 1414, 1424) are of dolerite and basalt veins, none of which, however, shows the poikilo-plektophitic texture of Harris's group B3. The specimens from Reviggerly are highly zeolitised with feldspar phenocrysts partially replaced by analcite, and the groundmass almost completely altered to analcite and fibrous zeolite or clay mineral. This alteration is probably deuteric. NI 1408, an 8-cm sheet 2 m below the top of the sill, is a fairly fresh sub-ophitic olivine-dolerite, with some interstitial zeolite and cracks infilled with limonite and goethite. The other specimens from Large Skerries are all thin vertical veins at the top of the intrusion and are notable for the included patches of very fine-grained material containing tabular feldspar, goethite, magnetite and granular pyroxene.

Coarse-grained pegmatite veins and sheets (Harris's group B4) are common in the upper part of the sill. A vein from near the boat-slip at Reviggerly consists of anhedra of brownish pyroxene, euhedra of labradorite fingered and meshed by analcite, abundant magnetite, often skeletal, interstitial analcite, and clumps of radiating natrolite needles which crystallised before the analcite. There is also some interstitial albite, with 'myrmekite-like' margins against limonite-like iron oxides. A vein from Castle Island (NI 1433) is similar in composition to the dark mafic clots from the central zone of the sill, consisting of partly altered olivine and labradorite with subordinate augite and magnetite, while an adjacent vein (NI 1436) is similar to the granophyric matrix of the central zone.

A pegmatitic vein from Winkle Island (NI 1440) has large feldspar, subophitic pyroxene, and skeletal iron ore grains in a matrix almost entirely replaced by fibrous zeolites which also interfinger with the feldspars. The matrix contains small tabular crystals of sodic plagioclase.

Harris (pp. 124–6) describes porphyritic leucocratic veins with pegmatitic margins. These may be matched by a 25-mm vein from Castle Island (NI 1454) which has a core of fine-grained basalt, with abundant granular augite, copious magnetite octahedra, and tabular and lath-shaped microphenocrysts of labradorite, sometimes glomerophyric. The marginal zones, about 6 mm wide, are of zeolitised granophyric dolerite, the contact between the centre and margin being marked by a concentration of magnetite and dolerite.

The islands of the Skerries have also yielded several composite veins of similar but more complicated type. For instance a narrow 13-mm vein from the upper part of the sill on Large Skerries, intruded into coarse olivine-dolerite, has the following composition (NI 1413):

1 A zone of fine-grained basalt consisting of granular augite, feldspar laths and abundant magnetite. This averages about 1 mm wide but is irregular and sends off apophyses into the dolerite. The contact with the dolerite is irregular but sharp.

2 A feldspar-rich zone 2 mm to 3 mm wide. Notably yellow in colour, it consists of laths of labradorite with granular augite and magnetite in a matrix of yellow-brown goethite.

3 An impersistent zone of very fine-grained microporphyritic basalt, 0–2 mm wide. Microphenocrysts of labradorite and augite which may be micro-xenoliths, in a groundmass of granular augite, labradorite microlites, and copious magnetite.

4 A zone 7 mm wide of medium-grained basalt with phenocrysts of augite and labradorite. The contact with zone 3 is irregular and sharp and the xenoliths in zone 3 are probably fragments of this zone. The groundmass of zone 4 is of small labradorite laths, minute granular augite and abundant magnetite. The junction with the dolerite is sharp and small fingers of basalt penetrate between the grains of the dolerite. The phenocrysts in this zone look like plucked-off fragments of the dolerite.

The largest vertical vein seen is at the east end of Castle Island where a massive sheet 0.6 m to 0.9 m wide can be followed for 30 m or more. The bulk of this vein is fine-grained but a coarse-grained marginal zone is present on one side or sometimes on both sides. The marginal zone (NI 1431) consists of analcitised olivine-dolerite with included fragments of porphyritic olivine-basalt. The basalt has phenocrysts of olivine, augite and labradorite in a matrix of granular augite, labradorite laths, and magnetite with a few large skeletal ophitic grains of augite, packed with feldspar laths but no magnetite. The central part of the vein (NI 1430), which makes up the bulk of its thickness, is apparently a tuff consisting of fragments of labradorite, augite, (broken euhedra), magnetite (broken octahedra), clumps of granular augite stained by yellow iron oxide, and fragmental quartz in a matrix of microcrystalline quartz, chlorite, and ?feldspar. It is very similar to intrusive tuffs from Rhum (Hughes, 1960).

A 40-mm-wide vein from Winkle Island (NI 1438) has irregular bands of different grain size, though all consist of labradorite and granular augite with magnetite inclusions. Granular magnetite is especially abundant in the finer bands. Olivine occurs sparsely. The bands in this vein appear to pass into each other and sharp contacts are unusual.

A large lenticular segregation sheet 1 m thick and about 2.5 m below the top of the sill on Large Skerries, itself has a layered structure. The upper half of the sheet consists of coarse ophitic pyroxenite with dark olivine-segregation clots. A middle portion weathering into thin laminae overlies a dark mafic base. The pyroxenite (NI 1418) consists of large augite plates, ophitically enclosing laths of labradorite. The augite is meshed by thin veinlets of a brown pyroxene of slightly higher refractive index, possibly reflecting slight resorption of the augite. The texture is sometimes 'granophyric' or 'runic' between augite and feldspar, perhaps the result of eutectic crystallisation. The central zone (NI 1419) is a coarse analcitised dolerite, with laths of labradorite and anhedral augite in a matrix of analcite and yellow fibrous zeolite. Patches of vermicular chlorite stained by yellow iron oxides occur interstitially and replacing the feldspar. Magnetite is surrounded by vermicular chlorite, and goethite or limonite is common. The basal zone, not always present, is a pyroxene-rich gabbro or pyroxenite like the upper zone, but without the olivine clots.

The hydrothermal veins of calcite–zeolite–chlorite described by Harris occur in crush-zones in the sill at the Wash Tub and Portscaddan. A true vein of zeolite, however, occurs at the west end of Large Skerries. The 13-mm vein (NI 1428) has been shown by X-ray examination to be a partially dehydrated laumonite (film X 2569). It is identical with material from Monte Catin, Tuscany, described as 'laumonite, var. caporcianite'. Though there is some confusion in the names of these minerals, the Skerries specimen is probably best described as leonhardite.

KNOCKSOGHEY SILL The sill is composed of olivine-dolerite and is typical of the normal relatively undifferentiated sills of Tertiary age common in north-east Ireland. The rock is ophitic, with titanaugite plates up to 3 mm across enclosing laths of labradorite (An_{55}). The latter are up to 5 mm long. Olivine, showing a little alteration to serpentine, occurs in two generations —phenocrysts up to 2.5 mm in diameter and smaller grains (0.1 mm) occurring in the mesostasis between augite plates (NI 2107). The olivine is zoned, with an iron-enriched outer rim, and the phenocrysts are richer in magnesia than the olivines in the groundmass (Tomkeieff and Patterson, 1947, p. 93). Magnetite amounts to about 4 per cent of the whole rock. The mesostasis consists of fibrous chlorite and a small amount of zeolite.

The upper part of the sill (NI 2109) is finer in grain and less rich in olivine than the main mass, though it contains rather more magnetite. There is, however, no evidence of gravitational layering or segregation in the sill as a whole.

The veins of dolerite-pegmatite in the sill consist of pyroxene and plagioclase in idiomorphic grains, with a mesostasis of iddingsite, zeolite, and iron ore. The feldspar is rather more basic than that in the country rock and is andesine/labradorite in composition.
J R H HEW

Sills in the Port More Borehole

Of the eight intrusions intersected by the borehole three are probably sills, particularly the massive dolerite from 437.8 m to 660.7 m. This intrusion, and that from 340.9 to 361.3 m, are porphyritic olivine-dolerites, though variations in grain-size do occur.

Texture is directly related to grain size and ranges from variolitic in the finer-grained marginal facies (NI 3755, 3757, 3759, 3760) to ophitic (NI 3754, 3756, 3758, 3761–3). There is a gradual change from variolitic to ophitic texture as the general grain size increases, radiating clusters of plagioclase being present in nearly all cases. NI 3762 from a depth of 589.5 m shows a variant of the coarser textures producing a poikilophitic aspect with plates of augite about 0.75 mm across. The more normal ophitic texture is present in the same thin section.

Olivine crystals up to 2 mm long are universally present and form the main phenocrysts with the larger showing euhedral form. However, serpentinisation, initially around the margins and along planes more or less parallel to (001), modify the shape. Alteration along the good (010) cleavage occurs only at a late stage. Two sections, (NI 3754, 3762) exhibit two distinct generations of olivine, i.e. more or less euhedral phenocrysts and small ($\sim$0.04 to 0.06 mm) granular anhedra. In most cases the distinction is not obvious and, as the crystals rarely attain 2 mm in length, the gabbroic facies, NI 3763 (603.8 m), is non-porphyritic.

Included in the olivines, and less commonly in the groundmass are small (0.02 to 0.06 mm) brown or greenish-brown spinels. Those in the groundmass are usually rounded or, as in NI 3760, skeletal. The latter section shows a hollow square crystal 0.27 mm across with walls about 0.06 mm thick. The margins are irregularly rounded and the interior is filled with basaltic material. Spinels enclosed in olivine are usually perfectly euhedral.

Abundance of spinel may indicate a xenocrystic origin for much, if not all, the olivine. It was certainly formed prior to the emplacement of the magma as the grain-size variation is restricted and not dependent on that of the rock as a whole. Large crystals occur adjacent to the junction with the country rock (NI 3755) and the coarser rock is devoid of phenocrysts. While most crystals are rimmed by serpentine, section NI 3762 shows fresh euhedra with etched margins. Composition appears to be fairly uniform at about Fo_{85}.

Augite is the only pyroxene present, of neutral to pale green colour, occurring as anhedra moulded upon or intergrown with feldspar.

Plagioclase occurs mainly as laths although some platy crystals are present. The former may be curved and frequently form radiating bundles. Phenocrysts, up to 1 mm long, occur in the variolitic basalts where they frequently exhibit forked ends and glassy or chloritic cores. Their composition ranges across the andesine–labradorite fields from An_{35} to An_{50}. Primary crystals are labradorite, which may show zoning outwards to andesine in the broader individuals.

Accessory minerals are restricted to subhedral or, rarely, skeletal magnetite, and perhaps apatite in hair-like crystals 0.001 mm across (NI 3763). The latter occurs in the chloritic mesostasis and may penetrate into feldspar laths.

Alteration products are variably developed and are probably mostly deuteric. In places the rock is completely decomposed to an incoherent gravel. Zeolitisation of the mesostasis occurs and to some extent causes channelling in the plagioclase. Serpentinisation of olivine, which is complete in the variolitic basalts, has been mentioned above. A certain amount of marginal chloritisation occurs in the pyroxenes, merging into a sparse chloritic/zeolitic mesostasis.

Minute (0.3 mm) circular amygdales occur sporadically. These are lined with chlorite and filled with calcite (NI 3754) or more usually with radially arranged aggregates of feathery zeolites.

Apart from being less felsic, the upper sill, at 340.89 m to 361.29 m, does not differ radically from the main sill (437.80 m to 660.70 m). It is probable, therefore, that the igneous rocks were emplaced penecontemporaneously.

Compared with the Portrush Sill those at Port More contain no apparent segregation bands or felsic veins. In mineralogy the rocks also show differences. Olivine at Port More appears to have a higher Fo content and is usually associated with spinel, which has not been noted at Portrush. Plagioclase is somewhat more sodic. Zeolites are poorly developed in contrast with Portrush.

R W S

Black Rock

This probable intrusion is of coarse-grained olivine-basalt. The olivine occurs in two generations. Large phenocrysts, up to 5 mm across, are partially serpentinised, and small grains in the groundmass are sub-ophitic to feldspar laths. The pyroxene is a lilac titanaugite occurring in small grains, locally sub-ophitic, and the feldspar is a partially chloritised labradorite. The mesostasis is of iron oxides and chlorite with local analcite which also occurs in amygdales (NI 675–6).

Dykes

With only two exceptions all the dykes in the area are of olivine-basalt. The rock varies only in coarseness—marginal selvages and thin sheets are fine-grained—and consists typically of olivine phenocrysts, usually cracked and partially serpentinised in a groundmass of ophitic or sub-ophitic titanaugite, labradorite laths and iron ore. A mesostasis of chlorite and occasionally zeolite occurs in some specimens and green chlorite also occurs as a vesicle filling. Typical dyke-rocks are NI 1464, 1478.

Two dykes of tholeiitic basalt have been described. They are a thin irregular sheet from Portnaboe (NI 1457) and the massive dyke at Roveran Valley Head (NI 1472). The former consists of laths of labradorite and granular augite in a mesostasis of fibrous chlorite with magnetite grains. The latter has laths of basic andesine in a brown isotropic glassy mesostasis, speckled with iron ore. Pyroxene has probably been replaced by patches of chlorite which also occurs as filling in small (about 0.6 mm) amygdales.

J R H H E W

CHAPTER 18
Geophysical investigations

Very varied geophysical measurements have been made within the area of the Causeway Coast (7) Sheet. Surface and airborne geophysical studies have been carried out over both land and sea, and geophysical surveys have also been made in a deep borehole on land. Most of these geophysical studies have been associated with regional investigations of the land areas and continental shelf of Ireland, but several detailed surveys have also been carried out in the area.

The regional investigations have comprised magnetic and gravity surveys over land and sea, and shallow seismic surveys at sea. Detailed deep seismic surveys have been carried out on land, while the other detailed surveys have comprised laboratory palaeomagnetic investigations on oriented rock specimens, borehole logging, measurement of the physical properties of rock specimens, and heat flow studies.

Much of the information afforded by the geophysical investigations in the Causeway Coast district is of special relevance since rocks of the Antrim Lava Series cover most of the area, and there is little other evidence concerning the distribution of rocks below the lavas. Also, the geophysical surveys over the sea area provide the most continuous structural information available for that area.

GRAVITY SURVEYS

Cook and Murphy (1952) made the first gravity measurements in the Causeway Coast area in 1950, when they established three gravity stations within the one-inch Sheet area, in the course of a regional survey of the north of Ireland. A more intensive survey of the mainland area of the Causeway Coast sheet was made by geophysical staff of the Geological Survey of Great Britain (now part of the Institute of Geological Sciences) in 1959 in the course of a regional survey of the whole of Northern Ireland. Worden gravity meters were used to establish an irregular grid of stations in the area, averaging about one per square kilometre. No measurements were made on any of the small offshore islands. The results of these surveys have been published by the Geological Survey of Northern Ireland (1967).

In 1972, staff of the Marine Geophysics Unit of the Institute of Geological Sciences made a gravity survey of the sea area of the Coast Causeway sheet, during a regional geophysical survey of the continental shelf north of Ireland. A LaCoste and Romberg air/sea gravity meter, mounted in m.v. *Researcher*, was used to obtain an analogue record of gravity along one east–west and several north–south traverses within the One-inch Sheet area. Navigation was by Decca Main Chain 3B, combined with sextant fixes when possible. The latter regularly produced positions within 0.2 nautical miles of the corresponding Main Chain fixes. It was not possible to make measurements inshore because of the ship's draught.

Bouguer anomaly map

The results of the 1959 land survey and the 1972 seaborne survey are combined in Figure 27 as a map showing contours of Bouguer anomalies computed against the 1930 International Gravity Formula at sea level, referred to a gravity datum value of 981.2650 cm/s^2 at Pendulum House, Cambridge. For land stations the necessary height reductions were based on density values estimated as appropriate to the geological strata between each station and mean sea level. The density values used were those accepted for use throughout Northern Ireland (for sources, *see* Bullerwell, 1961a, p. 238):
Antrim Lava Series   2.85 g/cm^3
Chalk   2.62 g/cm^3
For sea stations a rock density of 2.65 g/cm^3 was used for the Bouguer correction.

Although the seaborne gravity meter used produces a continuous record of gravity variation, the corrections necessary were only calculated and applied for points at approximately ten-minute time-intervals along traverse lines. The survey off the Causeway Coast was made during a period of inclement weather, and it was found that for many stations the ship's motion was too great for the gravity data to be reliable. Only the acceptable gravity stations are shown on the Bouguer anomaly map. The accuracy of the Bouguer anomaly values at sea is thought to be about ±5 mGal, so only the 5-mGal contours are given there.

The Bouguer anomalies are dominated by the area of low gravity field in the south-east corner of the district. This is part of the gravity trough, discovered by Cook and Murphy (1952), extending from Dungiven in the south-west some 51 km to the north-east to the coast at Ballycastle, and presumably through Rathlin Island from the evidence of the 1959 survey (Bullerwell, 1961b). This gravity trough coincides with the Dungiven–Ballycastle syncline postulated by Wright (1919) on the basis of geological evidence. He believed that it was a syncline of Carboniferous strata, resting on Dalradian basement, beneath the thick cover of the Antrim Lava Series. Cook and Murphy (1952, p. 22) showed from the gravity gradients, in conjunction with measured rock densities, that the observed gravity could not be produced by Carboniferous rocks alone, and suggested that the trough was probably filled mainly with Triassic deposits as well as with some Permian and Carboniferous rocks. The Port More Borehole, which was later drilled near the centre of the gravity trough, confirmed the presence of thick Triassic and possibly Permian deposits (unbottomed at 1897 m depth). It is interesting to note that the 1972 marine gravity survey shows that the Dungiven–Ballycastle gravity trough does extend offshore to the north-east, beyond Rathlin Island, but only for another 13 km. The minimum Bouguer anomalies, in fact, occur around Rathlin Island. Thus it appears that the basin of low-density sediments producing the low Bouguer anomalies does not deepen appreciably offshore, and that the Port More Borehole is sited almost in the middle of this basin.

In the mainland area of the Causeway Coast sheet, west of the Dungiven–Ballycastle gravity trough, the anomaly values increase on to a broad area of relatively high gravity which approximately coincides with the Ardmore Anticline postulated by Wright (1919). In this area the average field of +13 to +14 mGal is much lower than that on the south-east side of the Dungiven–Ballycastle gravity trough, where the anomalies rise as high as +27 mGal. South-east of the gravity trough Dalradian basement rocks outcrop or directly underlie the Antrim Lava Series, so it must be inferred that on the north-west side, in the Portrush area, perhaps one to two thousand metres of sedimentary deposits underlie the lavas.

Offshore the anomaly values increase fairly consistently to the north-west, presumably reflecting thinning of the sedimentary

144 CHAPTER 18 GEOPHYSICAL INVESTIGATIONS

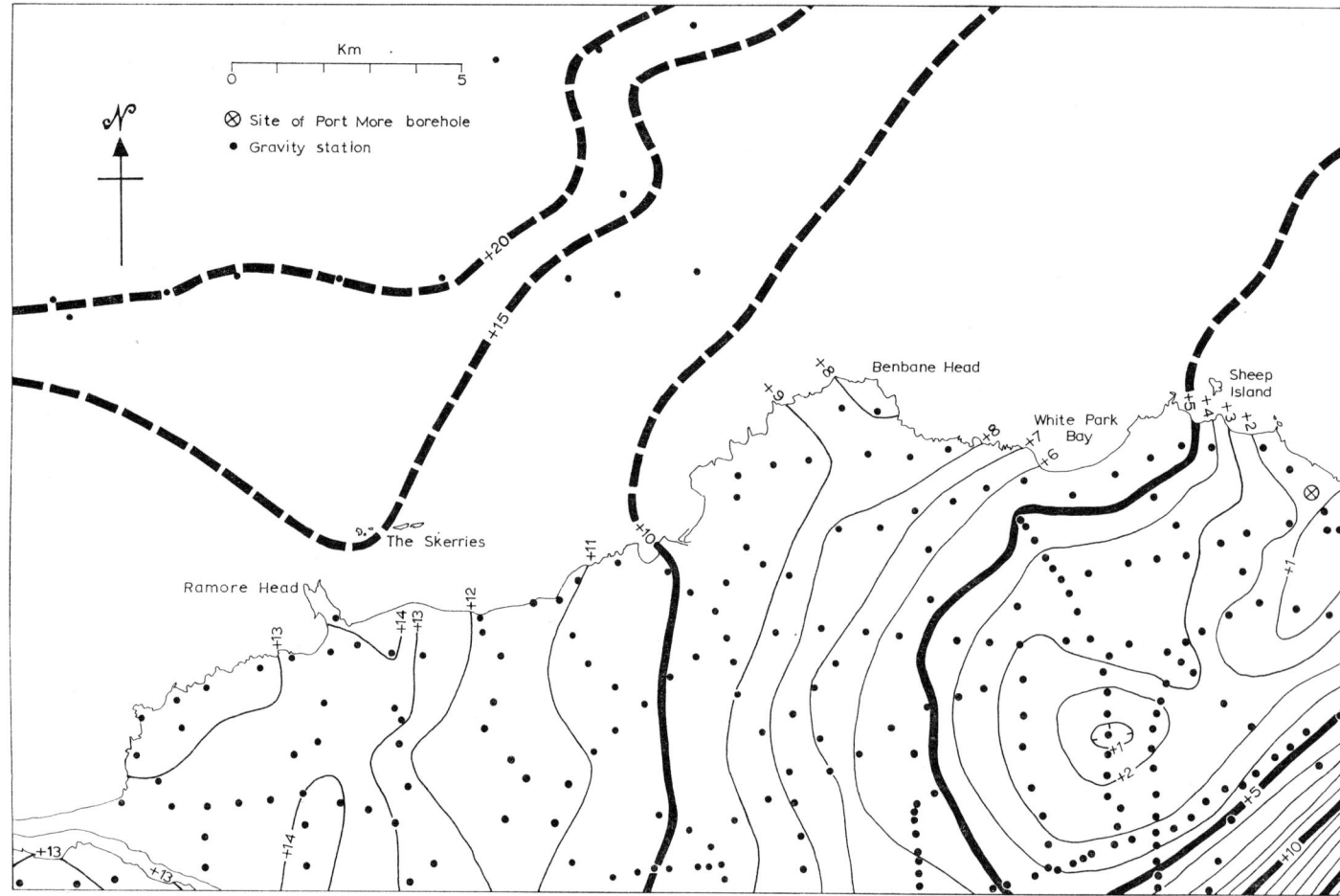

Figure 27 Map showing Bouguer gravity values at sea level. Contour interval is one milligal on land, five milligals at sea. Dots indicate gravity stations

rocks in that direction. The Lough Foyle gravity trough (Bullerwell, 1961b, p. 256; Murphy and others, 1971) does not appear to extend into this area.

MAGNETIC SURVEYS

Apart from the recognition of compass errors by mariners near to the north Irish Coast, the first magnetic surveys in the Causeway Coast area appear to have been made in 1834 by Lloyd, Sabine and Ross (1836). In the course of an investigation of the dip and intensity of the magnetic field throughout Ireland they established a station beside the River Bann, on the southern boundary of the One-inch Sheet, where they noted a 'local attraction apparently due to basalt' (p. 128). At this station they measured a magnetic dip of 71°22′ : their measurements of intensity were not absolute but calculated relative to that in London.

Palaeomagnetic studies

The Antrim Lava Series has been the subject of much palaeomagnetic investigation over a number of years, and many rock samples have been collected from the Causeway Coast area to be studied magnetically in laboratories. Hospers and Charlesworth (1954), who investigated the Lower Basalts of the Antrim Lava series, found that all their samples possessed a remanent magnetisation approximately opposite in direction to that of the present magnetic field. R. L. Wilson (1959) carried out a similar study of the Lower and Tholeiitic Basalts, as well as various intrusive bodies of Tertiary age, from the coasts of Antrim and Down, including the Causeway Coast area. Again he found the remanent magnetisation to be reversed and almost opposite to the present field: in addition, he observed little significant variation in the direction of magnetisation between the groups of samples of differing ages. In a later bipartite study, R. L. Wilson (1961) relied heavily on samples collected from the Causeway Coast sheet. In Part I he described a palaeomagnetic investigation of laterites, containing hematite and maghemite, forming the Interbasaltic Bed at the Giant's Causeway itself, where the laterites have been baked by overlying lava flows. Wilson established that the directions of magnetisation of the lava and baked laterite were the same, and that both were reversed. He was also able to calculate the strength of the ancient magnetic field as 42 microteslas (0.42 oe), compared to the present field of 53.5 microteslas (0.535 oe) in Northern Ireland. From the Curie point of hematite, and the conversion temperature of magnetite to hematite, he was also able to establish that the temperature of baking had been not less than 670°C and not more than 850°C, which may be relevant to petrological studies of basalts in the area. Wilson presented a large variety of evidence in Part II concerning reversal of the Earth's magnetic field during the Tertiary igneous activity in Northern Ireland. Included in this evidence was the similarity of magnetisation of the lavas and baked laterites from the Giant's Causeway, found in Part I, and also another investigation of samples from Portrush in the Causeway Coast area. He had determined the direction of magnetisation of sediments which have been strongly heated by the Portrush Sill (Tertiary dolerite), finding these sediments to be reversely magnetised, again in a direction agreeing closely with that indicated by various Tertiary igneous rocks. The evidence as a whole supports the reality of a reversed magnetic field in early Tertiary times, as opposed to self-reversal of the rocks in a normal field at some subsequent time in their history, and is among the most convincing of its type ever presented.

The results of the various palaeomagnetic investigations mentioned here have special significance in connection with the understanding and interpretation of magnetic force surveys in the Causeway Coast area, since they indicate that all the Tertiary igneous rocks sampled possess a strong remanent magnetisation opposite in direction to that of the present magnetic field. Hospers and Charlesworth (1954, p. 42) found the mean intensity of the natural permanent magnetisation of the Lower Basalts to be 1.7×10^{-3} cgs units. In addition, the range of rocks sampled is such that it can be assumed that most, if not all, the Tertiary igneous rocks of the area are reversely magnetised (see Bullerwell, 1966, p. 258).

Aeromagnetic survey

An aeromagnetic survey of the Causeway Coast sheet was made during 1959 as part of a survey of the whole of Northern Ireland and adjacent sea areas, arranged and supervised by the Geophysical Department of the Geological Survey of Great Britain for the Ministry of Commerce, Government of Northern Ireland. A Gulf magnetometer was used to make continuous recordings of total magnetic field during flights at a clearance of 1000 ft (305 m) above land or sea, along north–south traverses spaced at 2 km intervals and east–west tie lines at 10 km intervals. The results have been published as a map by the Geological Survey of Northern Ireland (1971).

Results obtained over the Causeway Coast sheet are shown in Figure 28: the contour values represent total force magnetic anomalies in nanoteslas (gammas) above a calculated linear regional field, which implies an increase by about 2.2 nT per kilometre northwards and about 0.3 nT per kilometre westwards. The anomalies over the mainland are intense and complex, with total amplitudes of up to 700 nT, due to the presence there of the basalt lavas. These complex anomales mask all other possible magnetic effects of strata below the basalts. The relatively low intensity of the anomalies over the sea areas, compared with those over the land, suggests that the lavas do not extend far offshore, except perhaps as far as the WSW–ENE magnetic high north-west of Ramore Head. It has already been stated by Bullerwell (1966, p. 259) that the anomaly configuration along the northern boundaries of the Antrim Lava Series usually consists of a positive anomaly maximum roughly along the boundary, with negative anomalies over the basalts, due to their reversed remanent magnetisation. This configuration is well illustrated in the Causeway Coast area by the series of intense WSW–ENE positive anomalies extending in echelon across the sheet, representing the northern boundaries of the basalts at different localities.

There is a strong correlation in the south-east corner of the One-inch Sheet between the aeromagnetic anomaly trend and the gravity gradient (Figure 27) across the Tow Valley Fault.

There is no apparent correlation between the magnetic and gravity anomalies offshore, where the magnetic anomalies are weak and irregular.

Marine magnetic surveys

A seaborne magnetic survey carried out by Riddihough (1968) between 1963 and 1965 extended into the north-west corner of the Causeway Coast area. However, the anomalies observed did not differ significantly from those revealed by the earlier aeromagnetic survey.

In addition to the gravity survey mentioned earlier, the 1972 IGS marine geophysical survey included continuous recording of the magnetic field along all traverse lines. The results of this magnetic survey have not yet been compiled but will be published in due course.

SEISMIC SURVEYS

It has already been explained here that the evidence of the gravity surveys suggested the presence of a relatively deep sedimentary basin beneath the Antrim Lava Series, north-west of the Tow Valley Fault. The possibility existed that this basin would contain strata ranging from Carboniferous to Cretaceous in age, many of which could be of economic interest. It was thought that seismic surveys might provide more positive evidence concerning the strata below the lavas than had been obtained from the gravity and magnetic evidence. Such information could have helped to decide whether a borehole into these strata was worthwhile, as well as determining the optimum site for such a hole. It was therefore decided to carry out an experimental reflection seismic survey in the area, and the Geological Survey of Great Britain placed a contract for the survey with Seismograph Services Limited, who carried out the work in 1960. In addition to the main reflection survey, they also carried out some local refraction investigations on representative outcrops of the various strata which were expected to underlie the area of survey.

Reflection survey

The seismic equipment used throughout the investigation comprised 24 AAV reflection–refraction amplifiers coupled to a 24-trace variable area recording camera and HS-1 reflection geophones with a natural frequency of 20 Hz.

Initially trial shots were made, south of the Causeway Coast area near Ballybogy. Charges of various sizes were fired in holes drilled to varying depths for the purpose of determining the optimum shooting and recording procedures. Because of the high velocity-contrast between the drift deposits and basalt, the major part of the energy from shots in drift was found to be reflected back from the top of the basalt to the surface, with the result that no discernible reflections were recorded. It was therefore necessary to drill into the basalt and best results were obtained using a charge of 4.5 kg of Geophex placed about 11 m below the top of the basalt.

Using this procedure a traverse was established along the main Ballycastle to Coleraine road, which crosses the south-east corner of the Causeway Coast sheet, with shot-points every 402 m. However the basalt along this traverse line proved to be much harder than in the trial locality, with the result that it was not found possible to drill into the basalt to the necessary depths; the reflections deteriorated; the records showed no persistent reflection alignment, and no interpretation could be made.

Refraction investigations

In the course of the reflection traverse, near-surface horizontal velocities ranging from 3.7 to 5.5 km/s were derived for basalt, and a much lower velocity of 2.2 km/s was measured for the Interbasaltic Bed. Formation velocities derived from special profiles established on outcrops of Chalk and Lias in the Causeway Coast area are listed below together with other formation velocities derived elsewhere:

Figure 28 Map showing total-force magnetic anomalies from an aeromagnetic survey flown at 1000 ft (305 m) above terrain or sea. Contour values referred to an arbitrary datum and expressed in nanoteslas (gammas). Positive contours in solid, negative in broken line

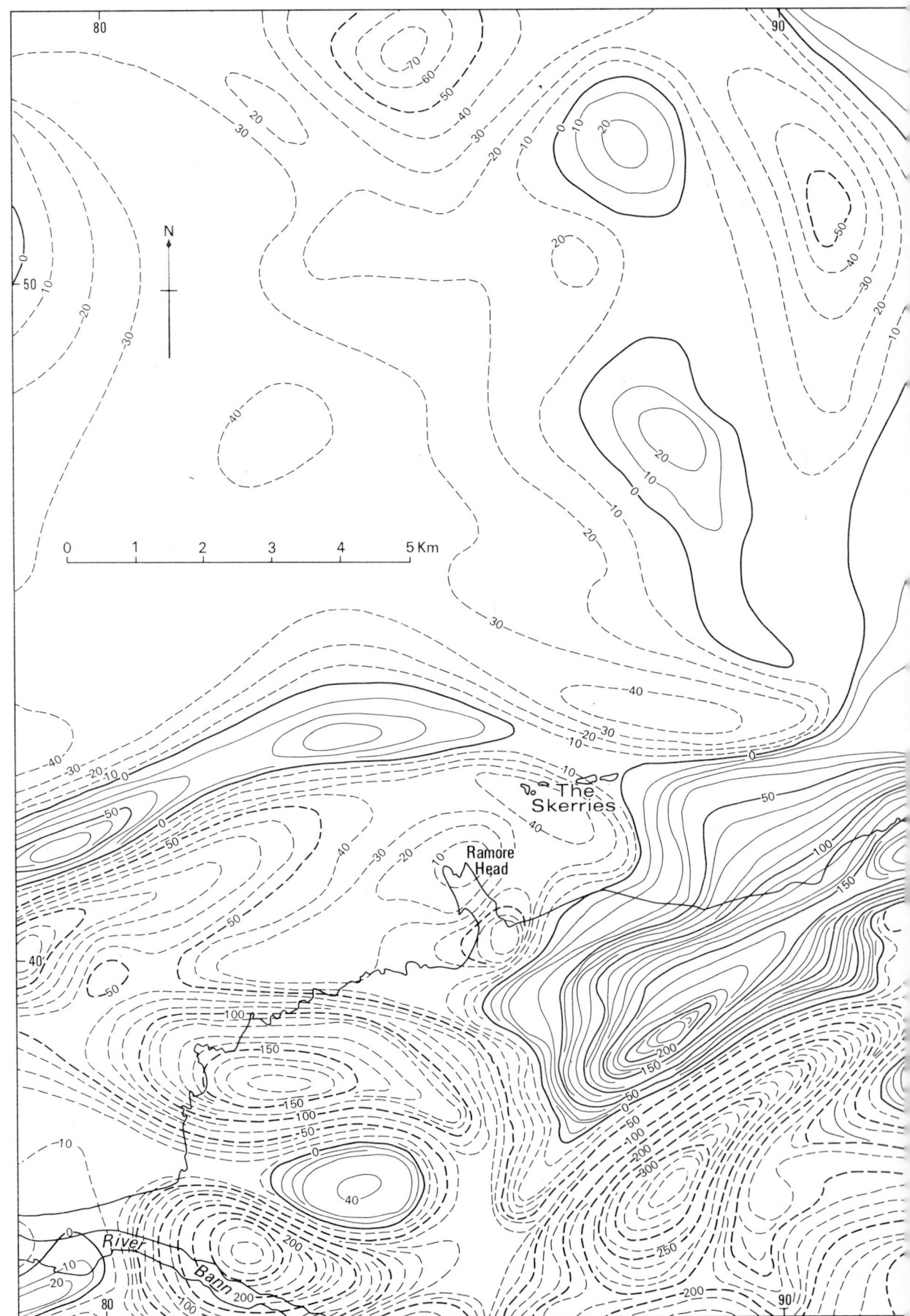

	Seismic velocity km/s	Locality
Antrim Lava Series: basalt	3.7–5.5	Ballycastle–Coleraine Road
Antrim Lava Series: Interbasaltic Bed	2.2	Ballycastle–Coleraine Road
Chalk	2.4	Portrush
Lias	4.5*	Portrush
Trias	2.3	Limavady
Carboniferous: Lower	2.6	Ballycastle
Dalradian	4.9	Ardgavan

* Obtained on a hornfelsed pavement so probably not typical

It is apparent from these figures that since the seismic velocities of the Mesozoic and Carboniferous strata are, in general, very much lower than that of the basalts, conventional refraction seismic methods could not be used to investigate the structure of the sedimentary basin in question, since the deep areas of the basin, as indicated by the gravity evidence, are almost entirely overlain by basalt. Also, it appears that the true velocity of the Chalk, from borehole measurements described in a later section of this chapter, is very much higher than that given above, and greater than the Mesozoic and Carboniferous velocities.

Marine seismic surveys

Continuous reflection seismic surveys were carried out along all traverse lines during the 1972 marine geophysical survey, by the Marine Geophysics Unit of IGS, using a 1000-joule multi-electrode spark source with a Geomecanique hydrophone and Huntec recorder. The results of these surveys have not been fully processed but will be published in due course.

PORT MORE BOREHOLE

The results of the regional gravity and aeromagnetic surveys of Northern Ireland were interpreted by Bullerwell (1964, p. 2) as 'indicating the possible existence at various localities in Northern Ireland of six sedimentary basins containing rocks of Upper Palaeozoic or Mesozoic age.' Any of these basins might contain coal and/or evaporite minerals of economic importance, so it was decided to drill a series of exploratory boreholes into the basins in turn. Following encouraging results in the first two of these holes at Larne, and at Magilligan (on the Lough Foyle gravity basin), the Port More borehole was sited in the area of low Bouguer gravity anomaly (Figure 27) on the Causeway Coast Sheet.

It was predicted on the basis of the gravity evidence that the top of the Triassic would be encountered at a depth of about 1500 ft (457 m) and the Carboniferous at about 5000 ft (1524 m). In fact Triassic rocks were first encountered at 660.7 m, but are immediately overlain by a dolerite sill 223.4 m thick which had not been expected, and the Permo-Triassic rocks had still not been bottomed when the hole was abandoned, due to technical difficulties, at 1896.7 m. A reassessment of the gravity evidence in relation to the drilling results is given at the end of this chapter.

At the outset of the drilling it was decided to carry out temperature surveys in the borehole during breaks in the drilling, in an attempt to determine the geothermal gradient at the site. Accordingly, attempts were made to lower borehole maximum thermometers mounted in a weighted sonde to the bottom of the hole each Sunday, when the temperature had had time to approach stability after the weekend cessation of drilling. It was hoped that in this way a series of bottom hole temperatures would be obtained at various depths, from which the temperature gradient could be calculated. However, it proved impossible to get the sonde to the bottom of the hole on each occasion. Since it is essential in such an investigation to have the thermometers adjacent to the hole bottom the main aim was not achieved, but the results obtained were sufficient to show that the temperature gradient was abnormally high. A request was therefore made to Mr J. Wheildon, of the Imperial College of Science and Technology, to carry out heat flow studies in the borehole after drilling had been completed. The borehole was kept open for this purpose and Mr Wheildon kindly complied with the request. The heat flow determined is one of the highest yet recorded in the British Isles and Mr Wheildon's account of the work is included later in this chapter.

Geophysical logs were made in the borehole by representatives of Schlumberger Ltd and the Atomic Energy Unit (now the Radioactive and Metalliferous Minerals Unit) of the Institute of Geological Sciences. The latter made gamma–gamma logs on two occasions, the first to 567.5 m and the second to 1033.3 m.

In 1970 Mr M. J. Bird of the Hydrogeological Department of the Institute of Geological Sciences made laboratory measurements on core samples of sandstones from the Sherwood Sandstone Group. His account of this work follows. JRPB

Core analysis

Effective porosity and intergranular permeability measurements were made on eight sandstone samples from the Sherwood Sandstone Group. These measurements were made in order to assess the magnitude of these properties at considerable depths in the Triassic, in the context of off-shore exploration for hydrocarbons.

EXPERIMENTAL METHODS AND APPARATUS From each of the sampled cores, right cylinders of 25-mm nominal length and diameter were cut in three directions. These plugs were taken vertically, and horizontally in two directions at 90° (HX and HY). In all 23 plugs were cut, there being only one horizontal plug for sample no. 672–8.

The plugs were tested in a gas permeameter similar to the type widely used in the petroleum industry, following the method described by the American Petroleum Institute (1956). The tests were carried out using air as the test fluid, assuming Darcy's Law for the flow of gases through porous media. The results were converted to equivalent values of permeability to a non-reactive liquid using Klinkenberg's correction data (Klinkenberg, 1941). The values were reproducible within ± 7 per cent in this investigation, where only one operator was used, but it is found that if measurements are made by two different operators the error is of the order of ± 14 per cent.

After the permeability determinations the plugs were tested for effective porosity using the liquid resaturation method described by the American Petroleum Institute (1960). They were dried in an oven overnight at 100°C, left to cool at room temperature, and then weighed individually. After weighing, they were placed in a dessicator where a vacuum was applied down to 5 torr (mmHg), before water was introduced to saturate the specimens, which were left to soak for 45 minutes. The vacuum was then released and the saturated plugs reweighed in both air and water, so that dry and saturated bulk density could be calculated, and hence effective porosity.

INTERPRETATION OF RESULTS The results of the measurements are listed in Table 4. The permeability values range from 1 millidarcy to 727 millidarcys, indicating rocks of a generally low order of permeability. In addition porosity can be seen to be of a low magnitude for the most part. The factors controlling porosity and permeability are grain size, sorting and cementation: grain size of the samples used is mainly small and sorting poor, with some sandstones showing a very wide range of grain size, thereby

Table 4 Laboratory determinations of density, porosity, permeability and isotropy of Sherwood Sandstone Group sandstones of the Port More Borehole

Sample number and code	Direction of plug	Depth m	Dry density g/cm³	Saturated density g/cm³	Effective porosity[3] per cent	Intergranular permeability millidarcys	Isotropy value
672–1V	Vertical		2.13	2.33	20	60	
672–1HX	Horizontal	1259.7	2.12	2.32	20	131	0.53
672–1HY	Horizontal		2.13	2.33	20	96	
672–2V	Vertical		2.16	2.35	19	105	
672–2HX	Horizontal	1350.3	2.20	2.39	19	298	0.25
672–2HY	Horizontal		2.14	2.33	19	532	
672–3V	Vertical		2.14	2.38	24	591	
672–3HX	Horizontal	1375.3	2.05	2.28	23	710	0.82
672–3HY	Horizontal		2.07	2.29	22	727	
672–4V	Vertical		2.23	2.39	16	103	
672–4HX	Horizontal	1556.0	2.23	2.39	16	109	0.42
672–4HY	Horizontal		2.18	2.36	18	378	
672–5V	Vertical		2.32	2.46	14	*	
672–5HX	Horizontal	1673.7	2.34	2.47	13	2	—
672–5HY	Horizontal		2.30	2.45	15	4	
672–6V	Vertical		2.41	2.52	11	<1	
672–6HX	Horizontal	1696.8	2.39	2.52	13	<1	—
672–6HY	Horizontal		2.40	2.52	12	<1	
672–7V	Vertical		2.17	2.36	19	76	
672–7HX	Horizontal	1800.5	2.16	2.35	19	100	0.76
672–7HY	Horizontal		2.15	2.34	19	*	
672–8V	Vertical		2.42	2.52	10	3	
672–8HX	Horizontal	1882.4	2.45	2.54	9	*	—

* Plug misshapen, so that permeability could not be tested: there is therefore no isotropy value.

having low porosity and pemeability values. Lack of cementation and uniform grain size results in the higher values of porosity and permeability of sample 672–3, whilst 672–8 has low porosity and permeability due to great variation in grain size and poor sorting. Low permeability values in 672–6 are caused by the occurrence of thin siltstone lenses. These have a great effect on permeability, owing to their irregular distribution, but only a slight effect in reducing porosity.

Comparison of the vertical and horizontal permeability values leads to a measurement of isotropy. This is a dimensionless parameter calculated by dividing the vertical permeability by the mean of the horizontal permeability values; when this value approaches unity the specimen is said to be isotropic.

Figure 29 shows the linear relationship between dry bulk density and porosity. There are four points which depart significantly from the linear curve—those relating to samples 672–3V, 672–2HX, 672–8V and 672–8HX. These unusual results are apparently caused by concentration of ferromagnesian minerals in 672–3V, the presence of siltstone lenses in 672–2HX, and the presence of feldspar and low density rock fragments in 672–8V and 672–8HX.

CONCLUSIONS The values of porosity and permeability measured are quite high when considered in relation to the depths from which the samples were taken. It is only in the bottom part of the sequence that porosity and permeability is very reduced. Discrepancies in the porosity–dry bulk density relationship occur at a series of horizons characterised by variable rock composition. Taken together the results indicate that part at least of the deeply buried Permo-Triassic sequence in this part of Northern Ireland is likely to be sufficiently permeable to have permitted movement of fluids towards potential hydrocarbon-bearing structures. M J B

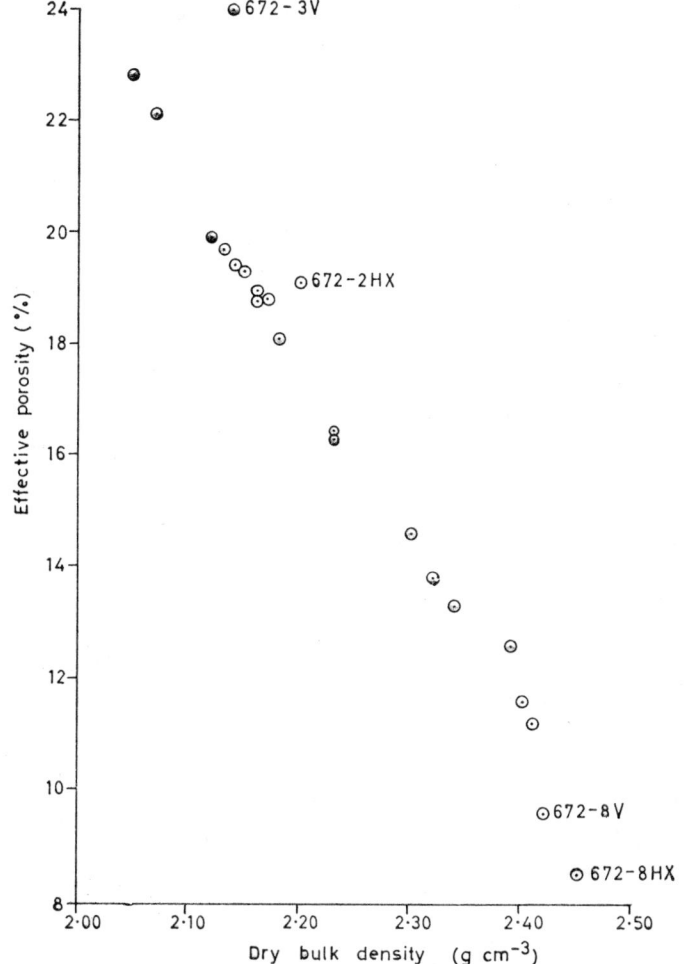

Figure 29 Graph showing relationship between dry bulk density and effective porosity of the Sherwood Sandstone Group

152 CHAPTER 18 GEOPHYSICAL INVESTIGATIONS

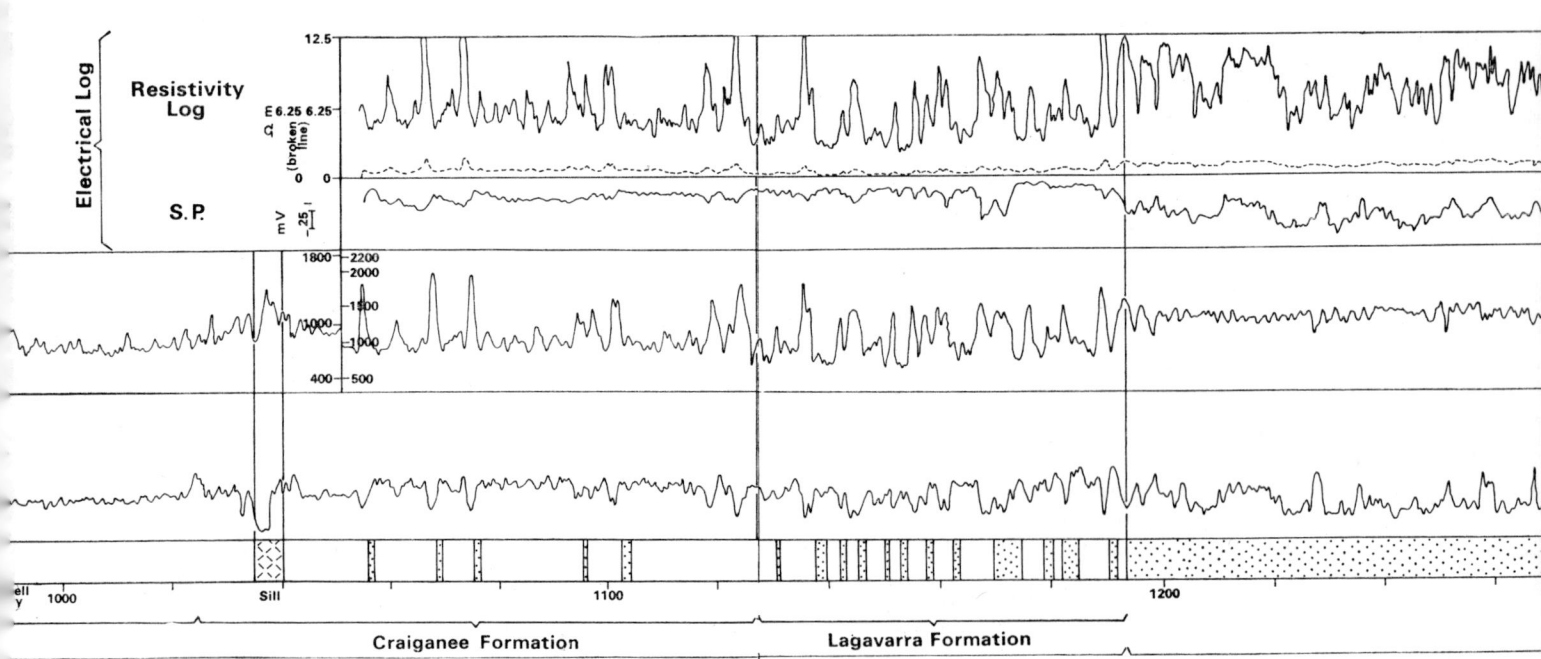

Sonic Log
km s⁻¹
Formation Density Log
g cm⁻³
Laterolog
Ω m (hybrid scale)
Neutron Log (run 1)
API units
Neutron Log (run 2)
API units
Gamma Ray Log (run 2)
µg Ra eq/ton
Lithology
METRES
Clay
100
200
Antrim Lava Series
White Limestone

Neutron Log
API units
Gamma Ray Log
µg Ra eq/ton
Lithology
METRES
800
Coolmaghera Skerry
900
Sixteen-inch Skerry
Sill
Knocksoghey Formation
Glenstaghey Formation
Mercia Mudstone Group

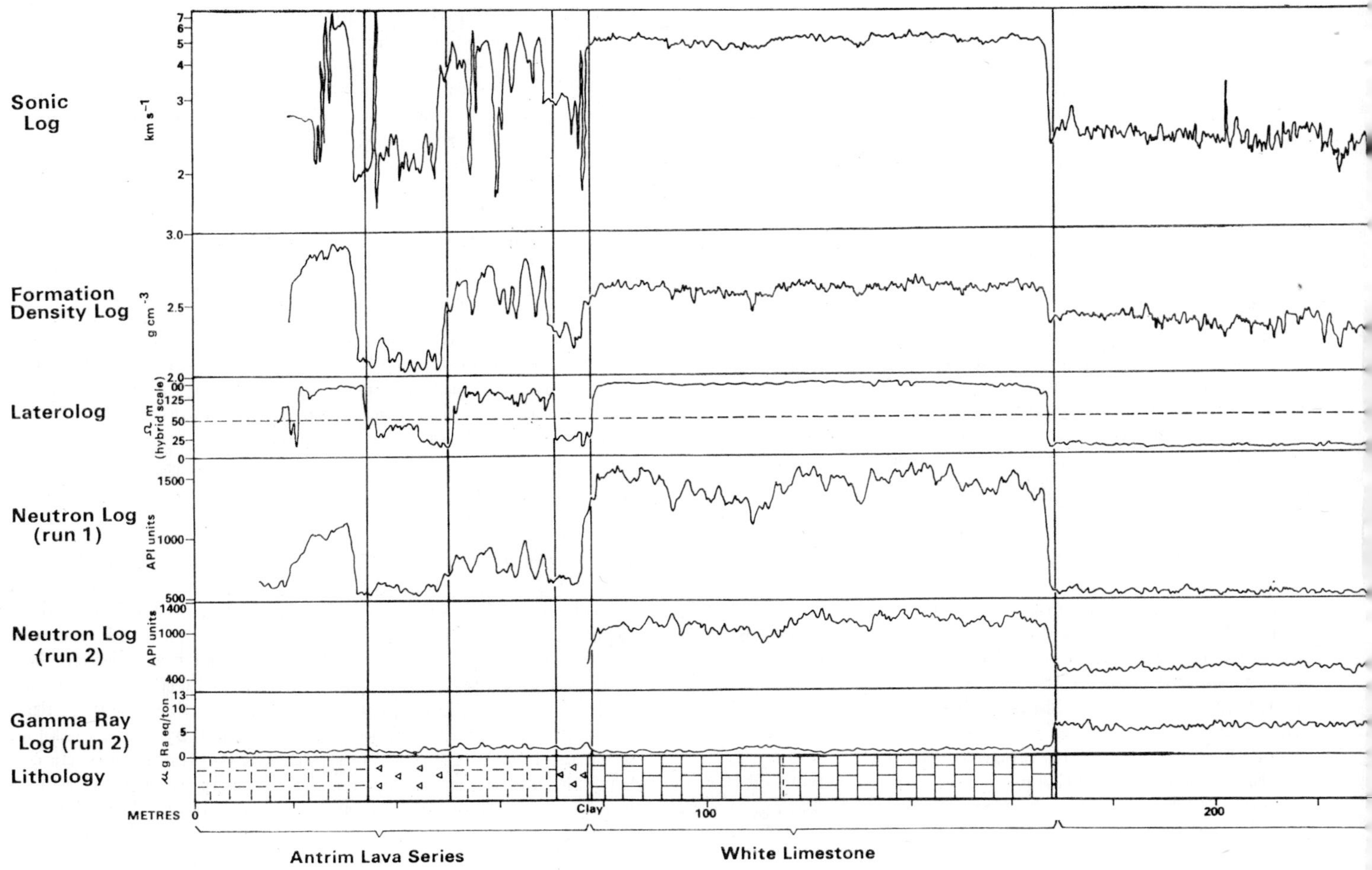

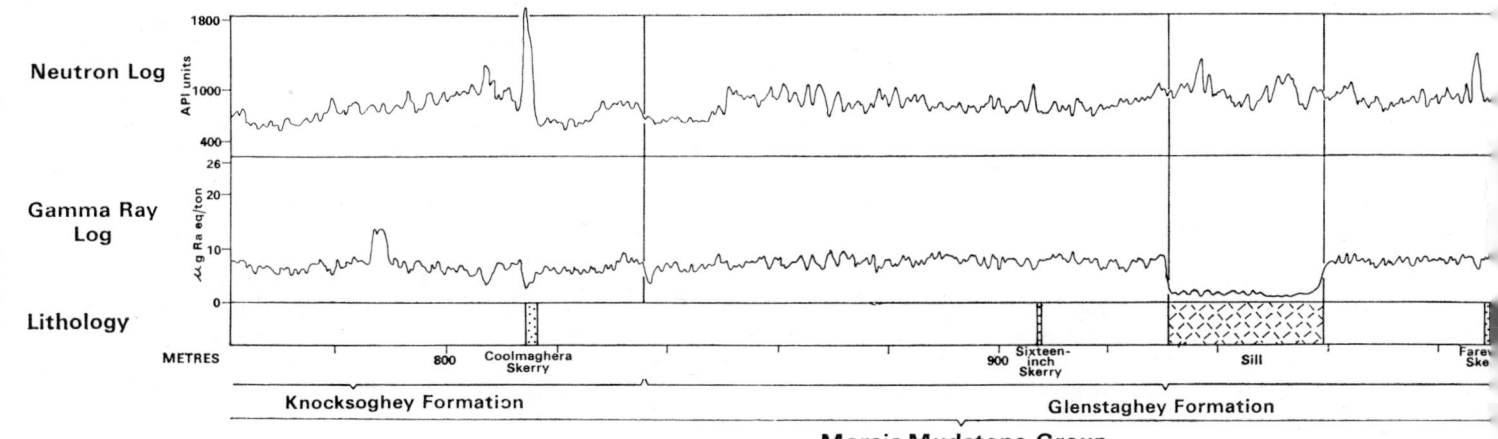

150 CHAPTER 18 GEOPHYSICAL INVESTIGATIONS

GEOPHYSICAL LOGGING

Geophysical surveys were made in the borehole by representatives of Schlumberger Ltd on two separate occasions, 26 June 1965 and 24 February 1967. At the time of the first survey the borehole had been drilled to a depth of 256 m and contained 14-in casing from the surface to 3.7 m and 9¼-in casing from the surface to 19.8 m, followed by uncased hole drilled with an 8 3/16-in bit. The following logs were recorded:
1 Temperature log, from surface to 255 m.
2 Standard electrical logs, comprising 16-in and 64-in Normal spacings, 18 ft 8 in Lateral spacing and Spontaneous Potential (SP) from 19.8 m (the casing shoe) to 255 m.
3 Micrologs, comprising 1 in × 1 in microinverse and 2 in micronormal spacings, and also a Microcaliper log measuring maximum borehole diameter, from 19.8 m to 255 m.
4 Laterolog, from 19.8 m to 254 m.
5 Gamma-ray log, from surface to 255 m.
6 Neutron log, from 15.2 m (the fluid level) to 255 m.
7 Sonic log, from 19.8 m to 255 m.
8 Formation-density log, from 19.8 m to 255 m.

When the second survey was made the borehole had been drilled to its final depth, but below 1486.8 m the diameter was too small for any Schlumberger sondes, and above that depth was large enough only for some special slim sondes. The hole contained 14-in casing from the surface to 3.7 m, 9¼-in casing from the surface to 261.8 m, 7-in casing from 175.6 m to 360.9 m, 6-in casing from 342.6 m to 586.7 m, and 5-in casing from 547.1 m to 1052.2 m (all driller's depths), followed by uncased hole drilled with a 4¼-in bit to 1486.8 m. The following logs were recorded:
1 Temperature log, from surface to 1481.9 m.
2 Electrical log, comprising a single Normal spacing of 10 cm, and Spontaneous Potential from 1055.8 m (the casing shoe as indicated by the geophysical logs) to 1481.9 m.
3 Gamma-ray log, from surface to 1484.4 m.
4 Neutron log, from 76.2 m (the fluid level) to 1484.4 m.

All the Schlumberger logs were recorded simultaneously at two depth scales, 1/200 and 1/500 on the first occasion, and 1/200 and 1/1000 on the second occasion. The original films are deposited in the records of the Applied Geophysics Unit, Institute of Geological Sciences.

The temperature logs recorded during this series of measurements at Port More cannot be used to estimate the steady geothermal gradient as they were made not long after mud circulation had ceased. Nor did they reveal any anomalies due to ingress of water, or variation in lithology in the borehole, in the latter case presumably because such anomalies did not have time to develop. The maximum temperature recorded, at 1481.9 m, was 63.3°C.

On the first logging occasion the SP log was rendered useless by interference from local DC surface currents, but this did not occur on the second survey. The resistivity log recorded on the second occasion was unusual in having only a single Normal spacing, as opposed to the two Normal and one Lateral spacings of the Standard Electrical log. However, the small (10 cm) spacing gives a very good vertical resolution of resistivity. The Laterolog Type 3, used on the first survey at Port More, also gives good vertical definition since it produces a current sheet only 30.5 cm thick.

No positive Microlog separations were observed opposite permeable formations so it has not been possible to derive porosity estimates from this log.

On the first survey at Port More the gamma-ray log was made with an SGD–F scintillation-counter, with time-constant set to 3 seconds, and a logging speed of 366 m per hour; while the neutron log was made at the same time with, of course, the same time-constant and logging speed, using GNAM equipment which has a radium–beryllium source and a spacing of 50.8 cm. For the second survey the two logs were again run together, using an SGD–H scintillation-counter and GNT–K neutron equipment, with an americium–beryllium source and a spacing of 40.6 cm. This equipment has a diameter of only 5 cm. The time-constant was 3 seconds and the logging speed 366 m per hour below 1051.6 m, changing to 2 seconds and 549 m per hour above 1051.6 m. Both gamma-ray logs were calibrated to read directly the true radioactivity of the formations only under certain standard conditions. The corrections necessary to allow for departure from the standard conditions have been included in the radioactivity values quoted in the text: however, the gamma-ray log reproduced in Figure 30 has not been corrected.

The sonic log at Port More was made with the borehole-compensated type of equipment, using two receivers with a separation of 61 cm. The formation-density log was also borehole-compensated and was made with a time-constant of 3 seconds and a logging speed of 366 m per hour.

Assessment of the geophysical logs
(Figure 30, pp. 151–153)

The temperature logs, which show no significant anomalies, are not shown in Figure 30: likewise, the Microlog has been omitted since it is off-scale in most of the formations where it would be expected to be of interest in this case. The standard electrical logs of the first survey are also omitted since the resistivity logs were not suitable for reproduction at a reduced scale and the SP log is unreliable: also, the electrical logs in this case yield little information over and above that provided by the Laterolog, which is shown. It should be noted that the Laterolog scale is a 'hybrid' one, being linear in resistivity only up to 50 Ω m, and linear in conductivity, the inverse of resistivity, above 50 Ω m. Both neutron logs are shown for completeness, since there is considerable overlap between them, and that obtained during the second survey is much affected by casing. There is little difference between the two gamma-ray logs, where they overlap, so only the complete one is shown.

The geophysical logs show that the physical properties of the Antrim Lava Series are very variable with depth. Not only is there a sharp contrast between the main lithological units of the Series, basalt and tuff, but also there are large variations within these units. Both the basalts and tuffs show quite uniformly low natural radioactivity levels of about 1.1 micrograms of radium equivalent per ton of formation (hereafter μg Ra eq. per ton). The neutron log, which responds inversely to the total amount of hydrogen in the formation (almost all the hydrogen is contained in the water held in the formation either interstitially, or as water of crystallisation in clay and evaporite minerals) and hence reflects porosity, shows a higher response in the basalts (somewhat variable in the lower flows) than in the tuffs, indicating that the basalts are less porous than the tuffs, as might be expected. The upper flows of basalt, down to 33.9 m, appear from the geophysical logs to be more massive than the lower layers, from 49.8 m to 70.5 m, since the neutron response, resistivity, bulk density, and sonic velocity are much less variable in the upper layers than in the lower ones. The boundaries between tuff and basalt are more precisely defined by resistivity on the Laterolog than on any of the other geophysical logs. The formation-density log shows that at Port More the density of the Antrim Lava Series attains the value of 2.85 g/cm^3, generally assumed for that formation for Bouguer gravity reductions, only above 32.6 m and that the density of tuffs within the Series can be as low as 2.05 g/cm^3. In fact, the average value, over 55.8 m of the Series logged, is only 2.45 g/cm^3, and although this may not be truly representative of the Series in Antrim, it would appear that the value generally accepted may well be excessive. The sonic velocity of the Antrim Lava Series at

PORT MORE BOREHOLE 153

Figure 30 Geophysical logs of the Port More Borehole

Note The scale at the right-hand end of the self-potential log should be used (the scale at the left-hand end is incorrect). Also, the scales for the broken line of the resistivity log should have the numbers 62.5, not 6.25, and those for the Laterolog should have ∞, not 00

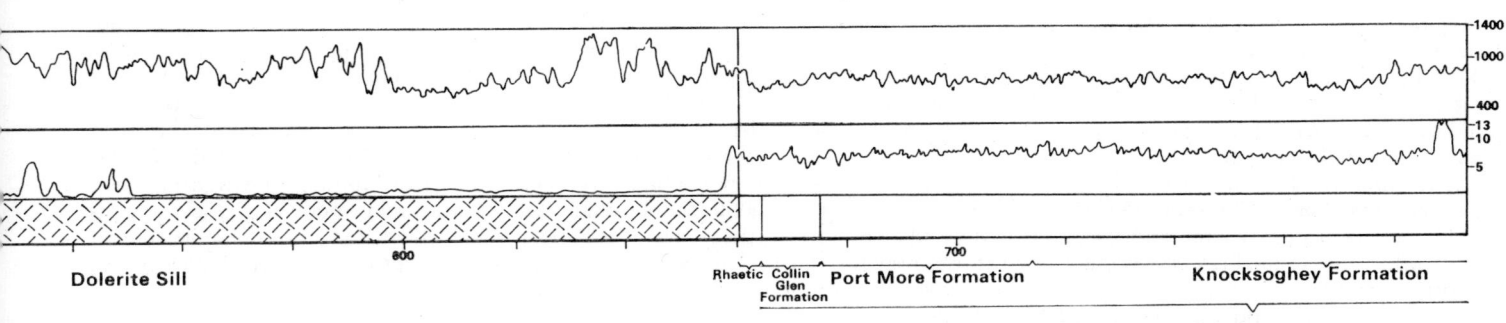

Port More is also extremely variable although it is suspected that the more extreme variations of very short wavelength in the log are instrumental in origin. The average velocity, 2.1 km/s, of the tuff layer from 33.9 m to 49.8 m is similar to that derived for the Interbasaltic Bed from surface refraction-seismic surveys. The velocities indicated in the basalt layers also agree fairly well with those derived from surface studies which indicated values ranging from 3.7 to 5.5 km/s.

On the other hand, the velocity indicated by the sonic log in the Chalk is very much higher than that derived from surface studies. The Chalk in the borehole has a remarkably consistent velocity averaging about 5.0 km/s over some 90.2 m, whereas a value of only 2.4 km/s was obtained at Ballycastle. It is considered that the borehole value is more likely to be representative of the Chalk in the area than that derived from the surface studies.

There is a marked change in the geophysical logs, with the exception of the gamma-ray log, at the top of the Chalk at 77.1 m. The density of the Chalk is fairly uniform and averages 2.59 g/cm^3, in good agreement with the value of 2.62 g/cm^3 generally used for Bouguer gravity reductions in County Antrim. The resistivity is extremely high, in fact almost infinite according to the Laterolog, and the neutron-indicated porosity is low. Quantitative comparison of the sonic, formation-density, and neutron logs suggests that the mean porosity of the Chalk is about 6 per cent, very similar to the value of nearly 5 per cent quoted by Cook and Murphy (1952, p. 9) on the basis of laboratory measurements on samples. The natural radioactivity of the Chalk is very low, indicating that its clay content is very small: the average value is about 0.7 µg Ra eq. per ton.

The top of the Lias at 168.35 m is very distinct on all the geophysical logs shown. The radioactivity level rises to an average of about 5 µg Ra eq. per ton and the resistivity falls to less than 10 Ω m, because of the high clay content; the neutron response is very low because of the water of crystallisation in the clay minerals. The mean formation-density of the 91.4 m of Lias logged is 2.40 g/cm^3, discounting the very low density deflections which correspond to deep caving of the borehole wall. The mean sonic velocity of the Lias mudstones is 2.3 km/s—very much lower than the value of 4.5 km/s derived by surface refraction studies at Portrush. However, the surface investigation was carried out on a hornfelsed pavement adjacent to the Portrush Sill, so the velocity determined there is probably not representative of the Lias generally (Bullerwell, 1966, p. 263).

The gamma-ray and neutron readings are very similar in the sections above and below the dolerite sill from 340.9 m to 361.3 m, and are also very uniform in both. This sill and the thicker one from 437.4 m to 660.7 m show a much lower radioactivity, with a background of about 1.2 µg Ra eq. per ton, and higher neutron response than the Lias mudstones. The boundaries of these sills are very distinct on the gamma-ray log but are consistently about 2.7 m higher than the depths given in the lithological log. The high radioactivity peaks, of up to 10 µg Ra eq. per ton at 528.5 m, are very unusual within an igneous rock, but xenoliths or rafts of Liassic mudstone occur at depths corresponding to the high radioactivities. There are no low neutron responses corresponding to the radiometric 'highs' but this could be explained by the water in the mudstones having been driven off by baking.

The top of the Rhaetic is very distinct on the gamma-ray log but again the change occurs 2.7 m higher than the depth shown on the lithological log. From 658.4 m to 714.8 m the radioactivity level is relatively uniform, with an average of about 7.5 µg Ra eq. per ton, in the Rhaetic, Collin Glen Formation, and Port More Formation. Over the same interval the neutron response is also relatively uniform but lower than the general level in the overlying sill.

There is more variation on the gamma-ray and neutron logs in the Knocksoghey Formation from 714.8 m to 836.7 m than in the overlying beds. The mean radioactivity level in the formation is again 7.5 µg Ra eq. per ton, but there is a distinct 'high' between 786.7 m and 789.7 m, with a maximum of 14 µg Ra eq. per ton, which may reflect the presence of 'reduction bands' mentioned in the detailed lithological log; there is no obvious neutron log anomaly associated with this radiometric feature. The very high feature on the neutron log, centred at 814.1 m with a maximum value of 1920 API units, and the associated low radioactivity on the gamma-ray log, correspond precisely with the Coolmaghera Skerry, which is a sandstone with some anhydrite and a dolomitic cement. Staff of the Engineering Geology Unit made physical tests on core specimens from a depth of about 815.9 m in the Coolmaghera Skerry. These tests comprised complete density determinations, hence effective porosity, and sonic velocity measurements. The data determined are listed in Table 5. The very low measured porosity explains the high neutron response, while the unusually high density and sonic velocity (for a sandstone) presumably reflects the anhydrite and/or dolomite content of the Skerry. The low radioactivity of the Skerry indicates that its clay content is low.

High neutron and low gamma-ray deflections are also associated with the Sixteen-Inch and Farewell Skerries at 907.6 m and 988.8 m respectively, although in neither case are the deflections comparable to that of the Coolmaghera Skerry. The Sixteen-Inch Skerry is described as sandstone with some anhydrite and the Farewell Skerry as sandstone and gypsum. Gypsum should produce a low neutron response because of its water of crystallisation, so it may be relevant that no dolomite was observed in the two lower Skerries.

The dolerite sill from 932.2 m to 960.1 m shows a similar low radioactivity level and above-average neutron response to those higher in the succession. The boundaries of this sill, as indicated by the gamma-ray log, agree precisely with those of the lithological log, but a discrepancy arises again lower down as the neutron and gamma-ray features associated with the Farewell Skerry and the dolerite sill between 1034.4 m and 1040.5 m appear to be higher than the lithological log suggests.

The background radioactivity level in the Craiganee Formation from 1024.4 m to 1126.6 m is significantly higher, at 8.2 µg Ra eq.

Table 5
Laboratory determinations of density, porosity and sonic velocity in some Triassic rocks of the Port More Borehole

Unit	Lithology	Depth	Saturated density	Dry density	Grain density	Effective porosity	Sonic velocity
		m	g/cm^3	g/cm^3	g/cm^3	per cent	km/s
MERCIA MUDSTONE							
Coolmaghera Skerry	sandstone	815.9	2.70	2.69	2.73	1	4.9
SHERWOOD SANDSTONE							
Upper Sandstone Formation	conglomerate	1332.0	2.62	2.58	2.84	4	5.5
Upper Sandstone Formation	sandstone	1453.5	2.52	2.43	2.68	10	4.0

per ton, than in the overlying Glenstaghey Formation where it is 5.6 μg Ra eq. per ton. The sandstones within the Craiganee Formation have low radioactivities and high resistivities, suggesting that they are relatively clean, so the corresponding high neutron responses, indicating very low porosity, are somewhat surprising. However, the relatively small SP deflections associated with the sandstones confirm that their permeability is low also, and the detailed lithological log mentions that they contain anhydrite.

The increased sandstone content of the Lagavarra Formation, from 1126.6 m to 1194.2 m accounts for the greater fluctuations on the gamma-ray and neutron logs than in the higher beds. The SP deflections associated with the sandstones are relatively small, with the exception of the sandstone unit from 1170.9 m to 1176.0 m in the lithological log (1167.7 m to 1172.3 m on the geophysical logs), which produces strong negative SP deflections of up to 50 mV, indicating good permeability in the top and bottom parts of the unit. The SP deflection opposite the central part of this sandstone is low, as is the resistivity, while the neutron response is low and the radioactivity high, indicating that this part is more clayey with low porosity and permeability. Higher resistivity (up to 10.5 Ω m) and lower radioactivity (as low as 3.4 μg Ra eq. per ton) correspond to the higher SP deflections, all indicating a relatively clean sandstone but, as with the Skerries higher in the hole, the neutron response, of up to 1780 API units, is surprisingly high. Assuming a sandstone matrix, this neutron response indicates a porosity of only 10 per cent, but again anhydrite was observed in this sandstone.

The topmost beds of the Sherwood Sandstone Group, from 1194 m to about 1280 m, show a relatively uniform neutron response averaging 1350 API units, which indicates a maximum possible average porosity of 21 per cent, assuming a sandstone matrix, but the actual value is probably lower because of the effect of the clay content of the formation on the neutron log. This clay content must be appreciable since the background radioactivity over the same depth-interval is 4.1 μg Ra eq. per ton and the average 5.6 μg Ra eq. per ton. The strong SP deflections, of up to 70 mV with an average of about 40 mV, over the same depth interval indicate that the permeability is relatively high. One sandstone sample, from 1259.7 m, was analysed in the laboratory of the Hydrogeological Department for porosity and permeability. The *effective* porosity measured was 20 per cent and the intergranular permeability was found to vary in different directions between 60 and 131 millidarcys (Table 4). It is not possible to compare this laboratory-derived porosity value directly with the geophysical logs, because of the discrepancy in depth between these logs and the lithological log. However, it is of the same order as the average porosity estimated from the neutron log, although the latter is a *total* porosity since the neutron log detects all the fluid in a rock, regardless of whether it can be extracted or not.

Between about 1280 m and 1310 m the Sherwood Sandstone Group has a much higher radioactivity averaging 7.7 μg Ra eq. per ton, indicating a higher clay content than in the beds above. Also the neutron response is more variable, the SP deflections are much lower, indicating decreased permeability, and the resistivity is higher, averaging 11 Ω m as opposed to 8 Ω m between 1194 m and 1280 m.

Between 1310 m and the bottom of the geophysical logs the natural radioactivity level is relatively low, suggesting that the sandstones are much cleaner than those above, with the exception of several highly radioactive layers, notably those centered at 1452.1 m, where the maximum value is 21.0 μg Ra eq. per ton, and at 1482.5 m, where the reading went off-scale at 21.5 μg Ra eq. per ton. These anomalies occur at depths where the background level is about 5 μg Ra eq. per ton and are the highest radioactivities recorded in the borehole. There is nothing unusual recorded in the lithological log which would obviously account for the extremely high radioactivities: the only possible variation discerned in the detailed lithological description is the presence of small pebbles in the sandstones in the vicinity of the gamma-ray peaks, whereas sandstone clasts appear to be more usual elsewhere. There are no significant neutron, SP or resistivity features associated with the radiometric peaks. Physical tests were made by the Engineering Geology Unit on samples of a sandstone core from a depth of 1453.5 m in the borehole. Again it is difficult to relate geophysical log depths precisely to the drilling depths, but it was thought that this core would be from the radioactive band producing the gamma-ray anomaly centred at 4452.1 m. The results of the laboratory determinations are summarised in Table 5. Quantitative interpretation of the neutron log at the same depth suggests a total porosity of about 12 per cent, assuming a sandstone matrix. The porosity values derived from the laboratory measurements and neutron log are therefore closely comparable.

On the other hand, the conglomeratic layer between 1330.8 m and 1334.7 m in the lithological log (probably about 2 m higher in the geophysical logs) has a relatively low radioactivity, comparable to that of the surrounding sandstones, but produces a very high neutron response, with a maximum of 2550 API units, and has a much higher resistivity (maximum 36 Ω m) than the adjacent beds (average 10 Ω m). The neutron response indicates a maximum possible total porosity of only 3 to 4 per cent assuming a sandstone matrix, for the least porous part of the horizon. This is in good agreement with the results of laboratory determinations, by the Engineering Geology Unit, of the effective porosity of samples from a conglomerate core at 1332.0 m. These results are summarised in Table 5 along with density and sonic velocity measurements. The high sonic velocity of this conglomerate is probably caused by the high proportion of quartzite pebbles within it.

Ignoring the conglomeratic layer and highly radioactive layers already mentioned, the natural radioactivity, neutron response, and resistivity of the Sherwood Sandstone Group increase slightly with depth between about 1310 m and the base of the geophysical logs. The SP deflections, of up to nearly 90 mV, are generally high, and are larger than in the beds above. The fact that the radioactivities are generally lower below 1310 m than in the beds above suggests that the high SP deflections at depth indicate increased permeabilities rather than changes in connate water salinities. This conclusion is supported by the results of determinations of the permeability of sandstone samples carried out by the Hydrogeological Department (Table 4). The samples from 1350.3 m to 1375.3 m have much higher measured permeabilities, although again very variable in different directions, than that from 1259.7 m. Again it is not possible to compare precisely the average effective porosity values derived from the laboratory measurements with those derived from the neutron log, because of discrepancies in depths. However, the total porosity indicated by the neutron log is in the range of 14 to 20 percent within 2 m above or below 1350.3 m, and 14 to 18 per cent within 2 m above or below 1375.3 m, assuming a sandstone matrix in both cases. These values are somewhat lower than might be expected considering the average effective porosities of 19 per cent and 23 per cent respectively derived from laboratory measurements (Table 4).

JRPB

Heat-flow measurement

A determination of terrestrial heat flow requires two separate measurements: the rate of increase in temperature with depth (temperature gradient, r) and the thermal conductivity (k) of the rocks in which the temperatures are measured. Heat flow (q) is then calculated according to the formula $q = kr$.

Temperatures were measured in the Port More borehole using a thermistor resistance thermometer linked to a modified Wheatstone bridge. The precision of the measurements was better than 0.01°C. The temperatures were measured at 3 m intervals from the borehole collar to a depth of 579 m. To confirm that the borehole had returned to thermal equilibrium, temperature measurements were made at 12 months and again at 24 months after completion of drilling. The results are shown in Figure 31.

Thermal conductivities of small specimens of borehole core were measured in the laboratory by means of a divided-bar apparatus. The divided bar was calibrated against standards of fused and crystalline quartz. Standard deviations of measurement were usually less than 2 per cent, but the maximum accepted was 5 per cent.

As samples from only that section of the borehole in the dolerite sill from 437.4 m to 660.7 m were available for study, the heat-flow calculation is confined to the depth interval from 442 m to 579 m. Combining the temperature gradient and mean thermal conductivity of sixteen samples of dolerite from this section of the borehole yields a heat-flow value of 80.3 mW/m² (1.92 μ cal cm^{-2} s^{-1}). Corrections for the Pleistocene climatic variations increase this value to 87.0 mW/m² (2.08 μcal cm^{-2} s^{-1}). The effect of local topography was calculated and found to be negligible.

JW

CONCLUSIONS

The deep gravity trough into which the Port More borehole was drilled was found to be produced by a sedimentary basin containing low-density rocks, as predicted. The inferred depth to the top of the Triassic was rendered inaccurate by the presence of a thick dolerite sill at the base of the Jurassic, which was not apparent from the geophysical evidence. A sequence of low-density Triassic rocks was encountered, but of an even greater thickness than estimated from the gravity evidence. The discrepancy can be explained partly by the presence of numerous high-density dolerite sills revealed by the borehole; and partly by differences between some of the observed bulk densities of the various strata encountered in the borehole, and those which had been assumed for the purpose of the gravity reduction and interpretation. The latter were based on laboratory measurements on samples not all of which were from the Causeway Coast area. In particular, the average bulk density of the Triassic rocks may be somewhat higher than the value of 2.40 g/cm³ which is usually used in Ireland. Unfortunately no density log was obtained of the Triassic strata, but measurements have been made in the laboratory on samples from ten sandstone layers and one conglomerate layer (Tables 4–5). These cannot, of course, be accepted as typical of the Triassic sequence encountered, much of which is argillaceous, but considered in conjunction with the overall lithology of the Triassic, it would appear that the average bulk density of these strata at Port More must be close to 2.50 g/cm³. If this is so, and assuming that the post-Triassic sequence encountered at Port More remains uniform over the gravity trough, then the depth to the base of the Permo-Triassic at Port More may be as great as 3600 m, if there are no Carboniferous strata present. If Carboniferous strata contribute significantly to the gravity anomaly then the Permo-Triassic must be correspondingly thinner and the total depth of the basin considerably greater than 3600 m.

JRPB

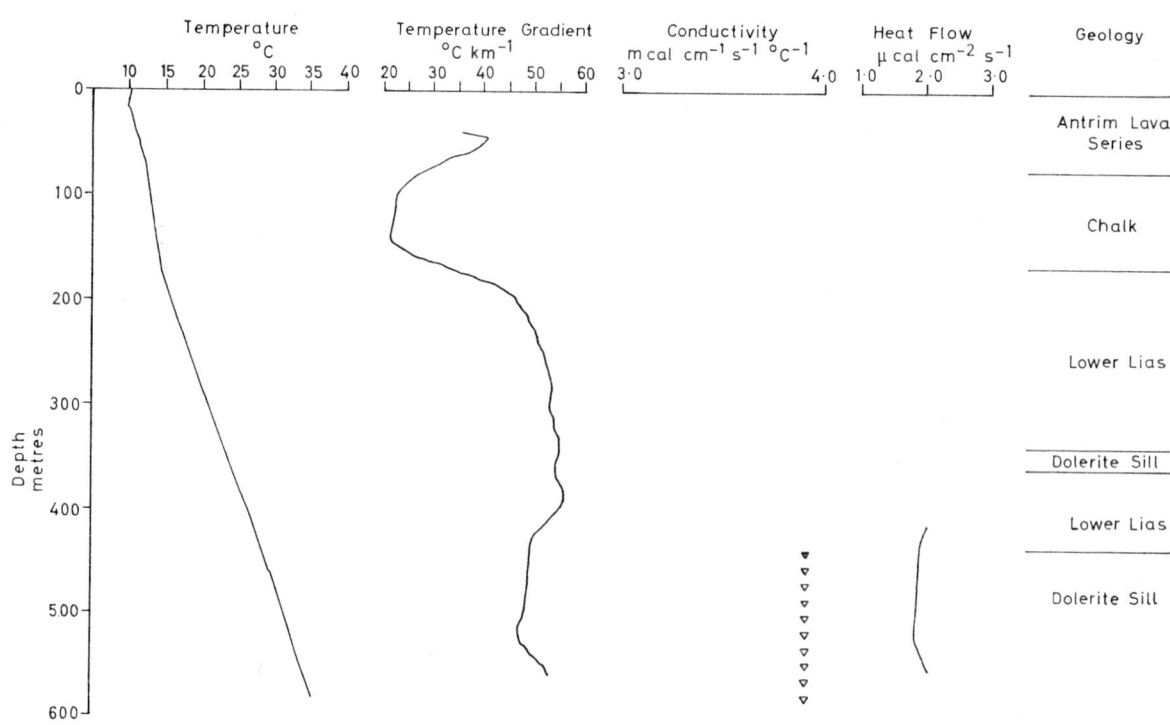

Figure 31 Correlation of borehole temperature measurements, laboratory measurement of conductivity, heat flow calculation and geology of the Port More Borehole above 600 m depth

BIBLIOGRAPHY

AMERICAN PETROLEUM INSTITUTE. 1956. Recommended practice for determining the permeability of porous media. Report No. 27. Dallas, Texas.
— 1960. Recommended practice for core analysis procedure. Report No. 40. Dallas, Texas.
ARKELL, W. J. 1933. *The Jurassic system in Great Britain.* 681 pp. (Oxford: Oxford University Press.)
BARRETT, W. F. 1890. Note on the magnetic properties of columnar basalt. *Sci. Proc. R. Dublin Soc.*, New Series, Vol. 6, pp. 382–383.
BELL, A. 1891. Notes upon the marine accumulations in Largo Bay, Fife, and at Portrush, County Antrim, North Ireland. *Proc. R. Phys. Soc. Edinburgh*, Vol. 10, pp. 290–297.
BERGER, J. F. 1816. On the geological features of the north-eastern counties of Ireland. *Trans. Geol. Soc. London*, 1st Ser., Vol. 3, pp. 121–195.
BERGSTRÖM, J., CHRISTENSEN, W. K., JOHANSSON, C. and NORLING, E. 1973. An extension of Upper Cretaceous rocks to the Swedish west coast at Särdal. *Bull. Geol. Soc. Denmark*, Vol. 22, pp. 83–154.
BROMLEY, R. G. 1967. Some observations on burrows of thalassinidean Crustacea in chalk hardgrounds. *Q. J. Geol. Soc. London*, Vol. 123, pp. 157–182.
BRYCE, J. 1834. An account of the celebrated Portrush Rock. *J. Geol. Soc. Dublin*, Vol. 2, pp. 166–174.
— 1835. On some caverns containing bones, near the Giant's Causeway. *Rep. Br. Assoc.*, [for 1834], pp. 658–660.
— 1837. On the geological structure of the north-eastern part of the County of Antrim. *Trans. Geol. Soc. London*, 2nd Series, Vol. 5, pp. 69–81.
BRYDONE, R. M. 1912. *The stratigraphy of the Chalk of Hants.* (London)
— 1915. The *Marsupites* Chalk of Brighton. *Geol. Mag.*, Decade 6, Vol. 2, pp. 12–15.
— 1933. The zone of granulated *Actinocamax* in East Anglia. *Trans. Norfolk & Norwich Nat. Soc.*, Vol. 13, pp. 285–293.
BULKELEY, R. B. 1693. A letter concerning the Giant's Causeway in the County of Antrim, in the north of Ireland. *Trans. R. Soc.*, Vol. 17, pp. 708–710.
BULLERWELL, W. 1961a. in FOWLER, A. and ROBBIE, J. A. 1961. Geology of the country around Dungannon. *Mem. Geol. Surv. North. Irel.*
— 1961b. The gravity map of Northern Ireland. *Ir. Nat. J.*, Vol. 13, No. 11, pp. 254–257.
— 1964. *Geophysical and drilling exploration in Northern Ireland.* H.M.S.O.
— 1966. in WILSON, H. E. and ROBBIE, J. A. Geology of the country around Ballycastle. *Mem. Geol. Surv. North. Irel.*
CHARLESWORTH, J. K. 1936. Large boulder of Ailsa Craig microgranite at White Park Bay, Co. Antrim. *Ir. Nat. J.*, Vol. 6, pp. 94–95.
— 1939. Observations on the glaciation of north-east Ireland. *Proc. R. Ir. Acad.*, Vol. 45, Sect. B, pp. 255–295.
— 1953. *The geology of Ireland.* 276 pp. (Edinburgh and London: Oliver and Boyd.)
— 1963. *Historical geology of Ireland.* 565 pp. (Edinburgh: Oliver and Boyd.)
— and PRESTON, J. 1960. *Geology around the university towns: North-east Ireland—The Belfast area. Geologists' Association Guide*, No. 18. 30 pp. (Colchester: Benham.)
— and others. 1935. The geology of north-east Ireland. *Proc. Geol. Assoc.*, Vol. 46, pp. 441–486.
— and others. 1960. The geology of north-east Ireland. *Proc. Geol. Assoc.*, Vol. 71, pp. 429–459.
CHRISTEN, M. 1902. Investigations into the glacial drifts of the north-east of Ireland. *Ir. Nat.*, Vol. 11, pp. 275–276.
CHRISTENSEN, W. K., ERNST, G., SCHMID, F., SCHULZ, M.-G. and WOOD, C. J. 1973. Comment on proposals for the designation of a neotype for the Upper Cretaceous *Belemnites mucronatus* LINK, 1807. Z.N. (S). 1160. *Geol. Jahrb.*, Reihe A, Vol. 9, pp. 41–45.
— — — — — — 1975. *Belemnitella mucronata mucronata* (SCHLOTTHEIM, 1813) from the Upper Campanian: Neotype, biometry, comparisons and biostratigraphy. *Geol. Jahrb.*, Reihe A, Vol. 28, pp. 27–57.
CLELAND, A. 1938. Red flints. *Ir. Nat. J.*, Vol. 11, pp. 275–276.
CLOOS, H. 1941. Bau und Tätigkeit von Tuffschloten: Untersuchungen an dem Swabischen Vulcan. [Structure and activity of tuff vents: investigations of the Swabian volcano]. *Geol. Rundsch.*, Band 32, pp. 703–800. [In German]
COFFEY, G. and PRAEGER, R.L. 1904. The Larne raised beach: a contribution to the Neolithic history of the north of Ireland. *Proc. Roy. Ir. Acad.*, Vol. 25, Section C, pp. 145–200.
COLE, A. J. 1895. The destruction of the Chalk. *Geol. Mag.*, Decade 4, Vol. 2, pp. 553–554.
— 1906. On contact phenomena at the junction of Lias and dolerite at Portrush. *Proc. R. Ir. Acad.*, Vol. 25, pp. 143–200.
— 1908. The red zone in the basaltic series of the County of Antrim. *Geol. Mag.*, Decade 5, Vol. 5, pp. 341–344.
— 1912. The Interbasaltic rocks (iron ores and bauxites) of north-east Ireland. *Mem. Geol. Surv. North. Irel.*
CONYBEARE, W. and BUCKLAND, W. 1816. Descriptive notes referring to the outline of sections presented by a part of the coasts of Antrim and Derry. *Trans. Geol. Soc. London*, 1st Series, Vol. 3, pp. 196–216.
COOK, A. H. and MURPHY, T. 1952. Measurements of gravity in Ireland. Gravity survey of Ireland north of the line Sligo–Dundalk. *Geophys. Mem. Dublin Inst. Adv. Stud.*, No. 2, Pt. 4.
CROOKSHANK, H. [Not dated]. *The true story of the Giant's Causeway.* (Dublin)
DEAN, W. T., DONOVAN, D. T. and HOWARTH, M. K. 1961. The Liassic ammonite zones and subzones of the north-west European province. *Bull. Br. Mus. Nat. Hist. (Geol)*, Vol. 4, No. 10, pp. 435–505.
DEMAREST, N. 1771. Sur l'origine et la nature du basalte à grandes colonnes polygonales, determinées par l'histoire naturelle de cette pierre, observée en Auvergne. [On the origin and nature of basalt with large polygonal columns, determined on the basis of the natural history of this rock observed in the Auvergne]. *Mem. Acad. Sci. Paris*, Vol. 87, pp. 705–775. [In French]

DONOVAN, D. T. 1966. The Lower Liassic ammonites *Neomicroceras* gen. nov. and *Paracymbites*. *Palaeontology*, Vol. 9, pp. 312–318.

DUNOYER, G. V. 1860. Notes on the stratigraphical position of the Giant's Causeway, and the structure of the basaltic cliffs immediately adjoining it. *The Geologist*, Vol. 3, pp. 3–14.

DUBORDIEU, J. 1812. *Statistical survey of the County of Antrim with observations on the means of improvement, drawn up by the order and for the consideration of the Dublin Society.* (Dublin)

DWERRYHOUSE, A. R. 1923. The glaciation of north-east Ireland. *Q. J. Geol. Soc. London*, Vol. 79, pp. 352–422.

EGAN, F. W. 1888. *See* SYMES, R. G., EGAN, F. W. and M'HENRY, A. 1888.

ELLIOTT, R. E. 1961. The stratigraphy of the Keuper series in southern Nottinghamshire. *Proc. Yorkshire Geol. Soc.*, Vol. 33, pp. 197–234.

EMELEUS, C. M. and PRESTON, J. 1969. Field excursion guide; the Tertiary volcanic rocks of Ireland. (Belfast)

ERDTMAN, G. 1928. Studies in the Postarctic history of the forests of north-western Europe. 1. Investigations in the British Isles. *Geol. Fören. Stockholm Förh.*, Band 50, pp. 123–192.

ERNST, G. 1963. Stratigraphische und gesteinschemische Untersuchungen im Santon und Campan von Lägerdorf (SW-Holstein). [Stratigraphic and geochemical investigations in the Santonian and Campanian of Lägerdorf (south-west Holstein)]. *Mitt. Geol. Staatsinst. Hamburg*, Heft 32, pp. 71–127. [In German]

— 1964. Ontogenie, Phylogenie und Stratigraphie der Belemnitengattung *Gonioteuthis* BAYLE aus dem nordwestdeutschen Santon/Campan. [Ontogeny, phylogeny and stratigraphy of the belemnite genus *Gonioteuthis* BAYLE in the Santonian/Campanian of north-west Germany]. *Fortschr. Geol. Rheinland & Westfalen*, Band 7, pp. 113–174. [In German]

— 1970. Zur Stammesgeschichte und stratigraphischen Bedeutung der Echiniden-Gattung *Micraster* in der nordwestdeutschen Oberkreide. [On the phylogeny and stratigraphical significance of the echinoid genus *Micraster* in the Upper Cretaceous of north-west Germany]. *Mitt. Geol.-Paläontol. Inst. Univ. Hamburg*, Vol. 39, pp. 117–135. [In German]

— 1971. Biometrische Untersuchungen über die Ontogenie und Phylogenie der *Offaster/Galeola*-Stammesreihe (Echin.) aus der nordwesteuropäischen Oberkreide. [Biometric investigations into the ontogeny and phylogeny of the *Offaster/Galeola* echinoid lineage in the Upper Cretaceous of north-west Europe]. *Neues Jahrb. Geol. Paläontol. Abh.*, Band 139, pp. 169–225. [In German]

— 1972. Grundfragen der Stammesgeschichte bei irregulären Echiniden der nordwesteuropäischen Oberkreide. [Basic questions of the phylogeny of the irregular echinoids in the Upper Cretaceous of north-west Europe]. *Geol. Jahrb.*, Reihe A4, pp. 63–175. [In German]

EVANS, E. E. 1966. *Prehistoric and Early Christian Ireland.* London.

EVANS, W. B. and WILSON, A. A. 1975. Outline of geology on Sheet 66 (Blackpool). 1:50 000. *Geol. Surv. G.B.*

EYLES, V. A. 1950. Note on the Interbasaltic horizon in Northern Ireland. *Q. J. Geol. Soc. London*, Vol. 106, pp. 136–137.

— 1952. The composition and origin of the Antrim laterites and bauxites. *Mem. Geol. Surv. North. Irel.*

FISHER, N. 1935. Locally extinct marine mollusca at Portstewart. *J. Conchol.*, Vol. 20, pp. 117–126.

FLETCHER, T. P. *In press.* The lithostratigraphy of Upper Chalk sediments in Northern Ireland. *Rep. Inst. Geol. Sci.*

FOLEY, S. 1694. An account of the Giant's Causeway in the North of Ireland. *Trans. R. Soc. London*, Vol. 18, pp. 170–172.

— 1694. Answers to Sir R. Bulkeley's queries relating to the Giant's Causeway, wrote down when we were upon the Causeway. *Trans. R. Soc. London*, Vol. 18, pp. 173–175.

FORBES, E. 1852. Figures and descriptions illustrative of British organic remains. Decade 4. *Mem. Geol. Surv. G.B.*

GARDENER, J. S. 1885. Lower Eocene plant-beds of the Basaltic Formation of Ulster. *Q. J. Geol. Soc. London*, Vol. 41, pp. 82–92.

GEOLOGICAL SURVEY OF NORTHERN IRELAND. 1967. Gravity anomaly map of Northern Ireland. 1:253 440.

— 1971. Magnetic anomaly map of Northern Ireland. 1:253 440.

GEIGER, M. E. and HOPPING, C. A. 1968. Triassic stratigraphy of the southern North Sea Basin. *Philos. Trans. R. Soc. London*, Series B, Vol. 254, pp. 1–36.

GEORGE, T. N. 1937. The geology of the district around Dunhampstead and Humbleton, Worcestershire. *Summ. Prog. Geol. Surv. G.B.* [for 1935], Pt. 2, pp. 119–129.

GETTY, T. A. 1973. A revision of the generic classification of the family Echioceratidae (Cephalopoda, Ammonoidea) (Lower Jurassic). *Paleont. Contrib. Univ. Kansas*, Pap. 63. 32 pp.

GRIBBON, H. D. 1969. *The history of water power in Ulster.* (Newton Abbot: David and Charles.)

HAMILTON, W. 1786. *Letters concerning the northern coast of the County of Antrim.* (London.)

HANCOCK, J. M. 1961. The Cretaceous System in Northern Ireland. *Q. J. Geol. Soc. London*, Vol. 117, pp. 11–36.

— 1972. Lexique Stratigraphique International. Vol. 1. Europe. Fasc. 3a. Angleterre, Pays de Galles, Ecosse. Fasc. 3a. XI. Crétacé. 162 pp. (Paris: Centre National de la Recherche Scientifique.)

HARKNESS, R. 1857. On the lignites of the Giant's Causeway and the Isle of Mull. *Rep. Br. Assoc.*, [for 1856], p. 66.

HARRIS, N. 1937. A petrological study of the Portrush sill and its veins. *Proc. R. Ir. Acad.*, Series B, Vol. 43, pp. 95–134.

HARTLEY, J. J. 1933. Notes on fossils recently obtained from the 'Chloritic' Conglomerate of Murlough Bay, Co. Antrim. *Ir. Nat. J.*, Vol. 4, pp. 238–240.

— 1938. On plant remains from the Interbasaltic Beds of Portrush. *Ir. Nat. J.*, Vol. 4, pp. 238–240.

HAWKES, J. R. and WILSON, H. E. 1975. The Portrush Sill, County Antrim, Northern Ireland. *Bull. Geol. Surv. G.B.*, No. 51, pp. 1–19.

HILL, A. R. and PRIOR, D. B. 1968. Directions of ice-movement in north-east Ireland. *Proc. R. Ir. Acad.*, Vol. 66, Section B, pp. 71–84.

HOLMES, A. 1936. A record of new analyses of Tertiary igneous rocks (Antrim and Staffa). *Proc. R. Ir. Acad.*, Vol. 43, Section B, pp. 89–94.

— 1945. The Giant's Causeway. *Nature*, Vol. 156, pp. 425–426.

HOSPERS, J. and CHARLESWORTH, H. A. K. 1954. The natural permanent magnetizaton of the Lower Basalts of Northern Ireland. *Monogr. Notes R. Astron. Soc., Geophys. Suppl.*, No. 7, pp. 32–43.

HOWARD, A. D. 1956. *Northern Ireland peat bog survey.* (Belfast.)

HUGHES, C. J. 1960. The Southern Mountains igneous complex, Isle of Rhum. *Q. J. Geol. Soc. London*, Vol. 116, pp. 111–138.

HULL, E. 1872. On the Raised Beach of the north-east of Ireland. *Rep. Br. Assoc.*, [for 1872], pp. 113–114.

HUME, W. F. 1897. The Cretaceous strata of County Antrim. *Q. J. Geol. Soc. London.*, Vol. 53, pp. 540–606.

IVIMEY-COOK, H. C. 1971. Stratigraphical palaeontology of the Lower Jurassic of the Llanbedr (Mochras Farm) Borehole. Pp. 87–92 in WOODLAND, A. W. (Editor). 1971. The Llanbedr (Mochras Farm) Borehole. *Rep. Inst. Geol. Sci.*, No. 71/18.

JACKSON, J. W. 1933–1938. Reports on excavations at the Ballintoy Caves, Co. Antrim. *Ir. Nat. J.*, Vol. 4, pp. 230–235; *Ibid.*, Vol. 5, pp. 104–109; *Ibid.*, Vol. 6, pp. 31–42; *Ibid.*, Vol. 7, pp. 107–112.

JEANS, C. V. 1968. The origin of the montmorillonite of the European Chalk with special reference to the Lower Chalk of England. *Clay Mineral.*, Vol. 7, pp. 311–329.

JELETZKY, J. A. 1951. Die Stratigraphie und Belemnitenfauna des Obercampan und Maastricht Westfalens, Nordwestdeutschlands und Dänemarks sowie einige allgemeine Gliederungs-Probleme der jüngeren borealen Oberkreide Eurasiens. [The stratigraphy and belemnite fauna of the Upper Campanian and Maastrichtian of Westphalia, north-west Germany and Denmark, together with some general problems of zonation in the boreal late Upper Cretaceous of Eurasia]. *Beih. Geol. Jahrb.*, Heft 1, pp. 1–142. [In German]

— 1955. Evolution of Santonian and Campanian *Belemnitella* and paleontological systematics: exemplified by *Belemnitella praecursor* Stolley. *J. Paleontol.*, Vol 29, pp. 478–509.

— 1964. *Belemnites mucronatus* Link, 1807 (Cephalopoda, Belemnitida): proposed designation of a Neotype under the Plenary Powers Z.N. (S.) 1160. *Bull. Zool. Nomencl.*, Vol. 21, pp. 268–296.

JELLET, J. H. 1879. Measurement of angles of basaltic columns in the Giant's Causeway, County of Antrim: made in 1877. *Mineral. Mag.*, Vol. 3, pp. 101–104.

JESSEN, K. 1949. Studies in late Quaternary deposits and flora-history of Ireland. *Proc. R. Ir. Acad.*, Vol. 52, Section B, pp. 85–290.

JUDD, J. W. 1886. On the gabbros, dolerites, and basalts, of Tertiary age, in Scotland and Ireland. *Q. J. Geol. Soc. London*, Vol. 42, pp. 49–97.

KELLY, J. 1868. On the geology of the County of Antrim with parts of the adjacent counties. *Proc. R. Ir. Acad.*, Vol. 10, pp. 235–327.

KHOSROVSCHAHIAN, R. 1972. Feinstratigraphische und faziesanalytische Untersuchungen im Campan von Misburg bei Hannover. [Investigations on the detailed stratigraphy and facies analysis in the Campanian of Misburg near Hanover]. Dissertation Braunschweig. [In German]

KILROE, J. R. 1888. Directions of ice-flow in the north of Ireland, as determined by the observations of the Geological Survey. *Q. J. Geol. Soc. London*, Vol. 44, pp. 827–833.

KIRWAN, R. 1793. Examination of supposed igneous origin of stony substances. *Trans. R. Ir. Acad.*, Vol. 5, pp. 51–81.

— 1799. *Geological Essays.* (London.)

KLAUS, W. 1964. Zur sporenstratigraphischen Einstufung von gipsführenden Schichten in Bohrungen. [On the stratigraphic classification of gypsiferous strata in boreholes by means of spores]. *Erdöl: Z. Bohrtech. Fordertech.*, Band 80, pp. 119–132. [In German]

KLINKENBERG, L. J. 1941. The permeability of porous media to liquids and gases. *Drill. Prod. Pract.*, pp. 200–213.

KNOWLES, W. J. 1899. Remains of the Great Auk from Whitepark Bay, County Antrim. *Ir. Nat.*, Vol. 8, pp. 4–6.

KONGIEL, R. 1962. On belemnites from Maastrichtian, Campanian and Santonian sediments in the Middle Vistula Valley (Central Poland). *Pr. Muz. Ziemi*, No. 5, pp. 1–148.

LA CROIX, A. 1886. Sur les roches basaltiques du Comté d'Antrim (Irlande). [On the basaltic rocks of County Antrim (Ireland).] *C. R. Acad. Sci. Paris*, Vol. 102, pp. 453–455. [In French]

LAMBERT, J. 1897. Note sur les échinides de la Craie de Ciply. [Note on the echinoids of the Ciply Chalk]. *Bull. Soc. Belge Géol. Paléontol. Hydrol.*, Vol. 11, pp. 141–190. [In French]

— 1903. Descriptions des échinides crétacés de la Belgique. 1. Étude monographique sur le genre *Echinocorys*. [Descriptions of the Cretaceous echinoids of Belgium. 1. Monographic study of the genus *Echinocorys*]. *Mém. Mus. R. Hist. Nat. Belg.*, No. 2, pp. 1–151. [In French]

— 1911. Descriptions des échinides crétacés de la Belgique. 2. Echinides de l'étage Sénonien. [Descriptions of the Cretaceous echinoids of Belgium. 2. Echinoids of the Senonian Stage]. *Mém. Mus. R. Hist. Nat. Belg.*, No. 4, pp. 1–81. [In French]

LAMONT, A. 1946. Red flints. *Ir. Nat. J.*, Vol. 8, pp. 398–399.

LANGTRY, G. 1875. On the occurrence of the Middle Lias at Ballycastle. *Rep. Br. Assoc.*, [for 1874], p. 88.

LLOYD, H., SABINE, E. and ROSS, J. C. 1836. Observations on the direction and intensity of the terrestrial magnetic force in Ireland. *Rep. Br. Assoc.*, [for 1835], pp. 117–162.

MACDONALD, G. A. 1953. Pahoehoe, aa, and block lava. *Am. J. Sci.*, Vol. 251, pp. 169–191.

MCGUGAN, A. 1957. Upper Cretaceous foraminifera from Northern Ireland. *J. Paleontol.*, Vol. 31, pp. 329–348.

— 1959. Upper Cretaceous foraminifera from Ballintoy, Co. Antrim. *Ir. Nat. J.*, Vol. 13, pp. 39–40.

— 1965. Liassic foraminifera from Whitepark Bay, County Antrim. *Ir. Nat. J.*, Vol. 15, pp. 85–87.

— 1974. Cretaceous foraminifera from Whitepark Bay, Northern Ireland. *Ir. Nat. J.*, Vol. 18, pp. 12–17.

MCGUIGAN, B. H. 1964. *The Giant's Causeway tramway.* (Lingfield.)

M'HENRY, A. 1888. See SYMES, R. G., EGAN, F. W. and M'HENRY, A. 1888.

— 1895. Sketch of the geology of County Antrim. *Proc. Geol. Assoc.*, Vol. 14, pp. 129–147.

MALLET, F. R. 1881. On the ferruginous beds associated with the basaltic rocks of north-east Ulster, in relation to Indian laterite. *Rec. Geol. Surv. India*, Vol. 14, pp. 139–148.

MANNING, P. I., ROBBIE, J. A. and WILSON, H. E. 1970. Geology of Belfast and the Lagan Valley. *Mem. Geol. Surv. North. Irel.* 242 pp.

— WILSON, H. E. and others. 1975. The stratigraphy of the Larne Borehole, County Antrim, Northern Ireland. *Bull. Geol. Surv. G.B.*, No. 50, pp. 1–50.

MARKHAM, R. A. D. 1971. The zones of the Gipping Valley Chalk, Suffolk. *Bull. Geol. Soc. Norfolk*, Vol. 20, pp. 26–28.

MAW, G. 1868. On the disposition of iron in variegated strata. *Q. J. Geol. Soc. London*, Vol. 24, pp. 351–400.

MAY, A. M'L. 1934. Excavations in caves at Port Braddon, County Antrim. *Ir. Nat. J.*, Vol. 5, pp. 56–58.

MELVILLE, R. V. 1956. The stratigraphical palaeontology, ammonites excluded, of the Stowell Park Borehole. *Bull. Geol. Surv. G.B.*, No. 11, pp. 67–139.

MITCHELL, G. F. 1951. Studies in Irish Quaternary deposits. *Proc. R. Ir. Acad.*, Vol. 53, Section B, pp. 111–206.

MOLYNEUX, T. 1694. Some notes on the foregoing account of the Giant's Causeway, serving to further illustrate the same. *Trans. R. Soc.*, Vol. 18, pp. 175–182.

— 1698. A letter containing some additional observations on the Giant's Causeway in Ireland. *Trans. R. Soc.*, Vol. 20, pp. 209–223.

Moody, G. T. 1905. The causes of variegation in Keuper Marl and other calcareous rocks. *Q. J. Geol. Soc. London*, Vol. 61, pp. 431–439.

Moskveena, M. M. 1959. *Atlas of the Upper Cretaceous fauna of the Caucasus and Crimea*. (Moscow.) [In Russian]

Murphy, T., Young, D. G. G. and Brück, P. M. 1971. The post-Dalradian strata along the north-west coast of Lough Foyle, Inishowen, Co. Donegal. *Proc. R. Ir. Acad.*, Vol. 71, Section B, pp. 171–181.

Naidin, D. P. 1960. The stratigraphy of the Upper Cretaceous of the Russian Platform. *Stockholm Contrib. Geol.*, Vol. 6, No. 4, pp. 39–61.

Nietsch, H. 1921. Die irregulären Echiniden der pommerschen Kreide. [The irregular echinoids of the Cretaceous of Pomerania]. *Abh. Geol.-Paläontol. Inst. Greifswald*, Band 2, pp. 1–47. [In German]

Nowak, J. 1913. Untersuchungen über die Cephalopoden der oberen Kreide in Polen. [Investigations on the cephalopods of the Upper Cretaceous in Poland]. *Bull. Int. Acad. Sci. Cracovie*, Series B, pp. 335–412.

Old, R. A. 1971. P. 48 in *Annu. Rep. Inst. Geol. Sci.* for 1970.

D'Orbigny, A. 1853–1860. *Paléontologie française. Description des mollusques et rayonnés fossiles. Terrains crétacés. 6. Echinoïdes irréguliers.* [Palaeontology of France. Description of the molluscs and fossil Radiata. Cretaceous formations. 6. Irregular echinoids]. (Paris.)

O'Reilly, J. P. 1879. Explanatory notes and discussion on the nature of the prismatic forms of a group of columnar basalts, Giant's Causeway. *Trans. R. Ir. Acad.*, Vol. 26, pp. 641–728.

— 1883. A cylindrical mass of basalt at Contham Head. *Proc. R. Ir. Acad.*, Vol. 2, p. 3.

Owen, E. F. 1970. A revision of the Brachiopod Subfamily Kingeninae Elliott. *Bull. Br. Mus. Nat. Hist. (Geol).*, Vol. 19, pp. 29–83.

Patterson, E. M. 1946. The Tertiary coal of Craigahulliar, Portrush. *Ir. Nat. J.*, Vol. 8, pp. 437–438.

— 1950. A preliminary note on the Tertiary lava succession in north Antrim. *Q. J. Geol. Soc. London*, Vol. 106, pp. 134–135.

— 1952. A petrochemical study of the Tertiary lavas of north-east Ireland. *Geochim. Cosmochim. Acta*, Vol. 2, pp. 283–299.

— 1955. The Tertiary lava succession in the northern part of the Antrim plateau. *Proc. R. Ir. Acad.*, Vol. 57, Section B, pp. 79–122.

— 1962. Tertiary vents in the northern part of the Antrim plateau. *Proc. Geol. Soc. London*, No. 1597, pp. 71–77.

— 1963. Tertiary vents in the northern part of the Antrim plateau, Ireland. *Q. J. Geol. Soc. London*, Vol. 119, pp. 419–443.

— and Swaine, D. J. 1955. A petrochemical study of Tertiary tholeiitic basalts: The Middle Lavas of the Antrim plateau. *Geochim. Cosmochim. Acta*, Vol. 8, pp. 173–181.

— — 1957. The Tertiary dolerite plugs of north-east Ireland. A survey of their geology and geochemistry. *Trans. R. Soc. Edinburgh*, Vol. 63, pp. 317–331.

Peake, N. B. 1972. *in* Hancock, J. M. 1972.

— and Hancock, J. M. 1961. The Upper Cretaceous of Norfolk. *Trans. Norfolk & Norwich Nat. Soc.*, Vol. 19, pp. 293–339.

— — 1970. The Upper Cretaceous of Norfolk. [Reprint with addenda and corrigenda] in *The Geology of Norfolk.*, edited by G. P. Larwood and B. M. Funnell. 107 pp. (London and Ashford, Kent: Headley Bros.)

Peck, R. E. 1973. *Applinocrinus*, a new genus of Cretaceous microcrinoids and its distribution in North America. *J. Paleontol.* Vol. 47, pp. 94–100.

Playfair, J. 1802. *Illustrations of the Huttonian theory of the Earth*. (Edinburgh.)

Pococke, R. 1748. An account of the Giant's Causeway in Ireland. *Trans. R. Soc.*, Vol. 45, pp. 124–127.

— 1753. A further account of the Giant's Causeway in the County of Antrim in Ireland. *Trans. R. Soc.*, Vol. 48, pp. 226–237.

— 1753. A letter upon the same subject with plate. *Trans. R. Soc.*, Vol. 48, p. 238.

Popiel-Barczyk, E. 1968. Upper Cretaceous terebratulids (Brachiopoda) from the Middle Vistula Gorge. *Pr. Muz. Ziemi*, No. 12, pp. 3–86.

Portlock, J. E. 1843. *Report on the geology of the County of Londonderry and of parts of Tyrone and Fermanagh*. (Dublin.)

Preston, J. 1962. Explosive volcanic activity in the Triassic sandstone of Scrabo Hill, Co. Down. *Ir. Nat. J.*, Vol. 14, pp. 45–51.

— 1963. The dolerite plug at Slemish, Co. Antrim, Ireland. *Liverpool & Manchester Geol. J.*, Vol. 3, pp. 301–314.

Prior, D. B. 1966. Late-Glacial and post-Glacial shorelines in north-east Antrim. *Ir. Geogr.*, Vol. 5, pp. 173–187.

Rao, T. V. M. 1928. A study of bauxite. *Mineral. Mag.*, Vol. 21, pp. 407–430.

Rasmussen, H. W. 1961. A monograph on the Cretaceous Crinoidea. *Biol. Skr. K. Dan. Vidensk. Selsk.*, Vol. 12, pp. 1–428.

Raspe, R. E. 1776. *An account of some German volcanoes and their productions with a new hypothesis of the prismatic basalts established upon facts, etc. Published as supplementary to Sir. W. Hamilton's observations on the Italian volcanoes*. (London.)

Reffay, A. 1969. Une côte à falaises basaltiques: le promontoire de la Chausée des Géants (Comté d'Antrim, Irlande du Nord). [A coast with basaltic cliffs; the promontory of the Giant's Causeway (Co. Antrim, Northern Ireland)]. *Rev. Géogr. Alpine*, Vol. 57, pp. 783–801. [In French]

— and Bouard, M. R. 1970. Contribution à l'étude des paléosols interbasaltiques à la Chaussée des Géants, Comté d'Antrim, Irlande du Nord. [Contribution to the study of the interbasaltic paleosols at the Giant's Causeway, Co. Antrim, Northern Ireland]. *Rev. Géogr. Alpine.*, Vol. 58, pp. 301–338. [In French]

Reid, R. E. H. 1958. Remarks on the Upper Cretaceous Hexactinellida of County Antrim. *Ir. Nat. J.*, Vol. 12, pp. 236–243, 261–269.

— 1968. Hexactinellid faunas in the Chalk of England and Ireland. *Geol. Mag.*, Vol. 105, pp. 15–22.

— 1971. The Cretaceous rocks of northern Ireland. *Ir. Nat. J.*, Vol. 17, pp. 105–129.

— 1973. The Chalk Sea. *Ir. Nat. J.*, Vol. 17, pp. 357–375.

Rhodes, P. S. 1958. *The Antrim coast road*. (Belfast: Northern Ireland Tourist Board).

Richardson, W. 1802. On basalts. *Philos. Mag.*, Vol. 13, pp. 129–136.

— 1803. Accounts of whynn dykes, in the neighbourhoods of the Giant's Causeway, Ballycastle and Belfast. *Trans. R. Ir. Acad.*, Vol. 9, pp. 21–43.

— 1803. Inquiry into the consisting of Dr. Hutton's theory of the Earth with the arrangements of the strata and other phenomena on the basaltic coast of Antrim. *Trans. R. Ir. Acad.*, Vol. 9, pp. 429–487.

— 1805. Remarks on the basalts of the coast of Antrim. *Trans. R. Soc. Edinburgh*, Vol. 5, pp. 15–20.

— 1806. On the volcanic theory. *Trans. R. Ir. Acad.*, Vol. 10, pp. 35–107.

— 1808. A letter on the alterations that have taken place in the structure of rocks on the surface of the basaltic country in the counties of Derry and Antrim. *Trans. R. Soc.*, Vol. 98, pp. 187–222.

Riddihough, R. P. 1968. Magnetic surveys off the north coast of Ireland. *Proc. R. Ir. Acad.*, Sect. B, Vol. 66, pp. 27–41.

Rittman, A. 1938. Die Vulkane am Myvatn in Nordost-Island. [The volcanoes at Myvatn in north-east Iceland]. *Bull. Volcanol.*, Série 2, Vol. 4, pp. 3–38. [In German]

Rohleder, H. P. T. 1926. A volcanic vent in the Whiterocks quarry, Portrush, N. Ireland. *Ir. Nat. J.*, Vol. 1, p. 98–100.

— 1927. Das Giants Causeway Gebiet (Co. Antrim in Nordirland). [The Giant's Causeway region, Co. Antrim, Northern Ireland]. *Z. Vulkanol.*, Vol. 10, pp. 232–267. [In German]

— 1928. A system of joints on the surface of basalt columns at the Giant's Causeway. *Geol. Mag.*, Vol. 65, pp. 354–355.

— 1929. *Geological guide to the Giant's Causeway and north coast of Antrim (between Portrush and Fair Head)*. (Belfast.)

— 1930. Der tertiäre Vulkanismus in Nord-Irland. [Tertiary volcanism in Northern Ireland]. *Geol. Charakterbilder*, Vol. 37. [In German]

Savage, R. J. G. 1963. Upper Lias ammonite from Cretaceous conglomerate of Murlough Bay. *Ir. Nat. J.*, Vol. 14, pp. 179–180.

Sandys, E. 1697. A true prospect of the Giant's Causeway. *Trans. R. Soc.*, Vol. 19, p. 233.

Scheuring, B. W. 1970. Palynologische und palynostratigraphische Untersuchungen des Keupers in Bölchentunnel (Solothurner Jura). [Palynological and palynostratigraphical investigations of the Keuper in the Bölchen tunnel (Solothurn Jura)]. *Schweiz. Paläontol. Abh.*, Vol. 88, pp. 1–119. [In German]

Schlüter, C. 1897. Über einige exocyclische Echiniden der baltischen Kreide und deren Bett. [On some irregular Echinoids of the Baltic Cretaceous and their horizon]. *Z. Dtsch. Geol. Ges.*, Band 49, pp. 18–50. [In German]

Schmid, F. 1967. Die Oberkreide-Stufen Campan und Maastricht in Limburg (Südniederlande, Nordostbelgien), bei Aachen und in Nordwestdeutschland. [The Campanian and Maastrichtian Stages of the Upper Cretaceous in Limburg (southern Holland, north-east Belgium), near Aachen, and in north-west Germany]. *Ber. Dtsch. Ges. Geol. Wiss.*, Reihe A, Band 12, pp. 471–478. [In German]

Scholle, P. A. 1974. Diagenesis of Upper Cretaceous chalks from England, Northern Ireland and the North Sea. Pp. 177–210 *in* Pelagic sediments: on land and under the sea. *Spec. Publ. Int. Assoc. Sedimentol.*, No. 1, pp. 177–210.

Simpson, J. B. 1961. The Tertiary pollen-flora of Mull and Ardnamurchan. *Trans. R. Soc. Edinburgh*, Vol. 64, pp. 421–468.

Smalley, I. J. 1966. Contraction-crack networks in basalt flows. *Geol. Mag.*, Vol. 103, pp. 110–114.

Smiser, J. S. 1935. A revision of the echinoid genus *Echinocorys* in the Senonian of Belgium. *Mém. Mus. R. Hist. Nat. Belg.*, No. 67, 51 pp.

— 1935. A monograph of the Belgian Cretaceous echinoids. *Mém. Mus. R. Hist. Nat. Belg.*, No. 68.

Smyth, T. 1870. On the geology of the coasts of Antrim and Londonderry, and on the age of the Giant's Causeway; being observations made in the north of Ireland in the autumns of 1865 and 1866. *Trans. Edinburgh Geol. Soc.*, Vol. 1, pp. 68–81.

Stephens, N. 1963. Late-Glacial sea levels in north-east Ireland. *Geography*, Vol. 4, pp. 345–359.

— and Synge, F. M. 1966. Pleistocene shorelines. Pp. 1–51 in *Geomorphological essays*, edited by G. H. Drury. (London: Heinemann.)

Stevens, G. R. 1973. Cretaceous belemnites. Pp. 385–401 in *Atlas of palaeobiogeography*, edited by A. Hallam. 531 pp. (Amsterdam, London and New York: Elsevier.)

Stewart, S. A. 1897. The Portrush raised beach. *Ir. Nat.*, Vol. 6, pp. 287–290.

Sweet, J. M. 1960. What is plinthite? *Mineral. Mag.*, Vol. 32, pp. 455–458.

Symes, R. G., Egan, F. W. and M'Henry, A. 1888. Explanatory Memoir to accompany Sheets 7 and 8. *Mem. Geol. Surv. Irel.*

Synge, F. M. and Stephens, N. 1966. Late- and post-Glacial shorelines and ice limits in Argyll and north-east Ulster. *Trans. Inst. Br. Geogr.*, Vol. 39, pp. 101–125.

Tate, R. 1865. On the correlation of the Cretaceous formations of the north-east of Ireland. *Q. J. Geol. Soc. London*, Vol. 21, pp. 15–44.

— 1867. On the Lower Lias of the north-east of Ireland. *Q. J. Geol. Soc. London.*, Vol. 23, pp. 297–305.

— 1870. Note on the Middle Lias in the north-east of Ireland. *Q. J. Geol. Soc. London*, Vol. 26, pp. 324–325.

— and Holden, J. S. 1870. On the iron ores associated with the basalts of the north-east of Ireland. *Q. J. Geol. Soc. London*, Vol. 26, pp. 151–165.

Thompson, J. A. 1907. Inclusions in some volcanic rocks. *Geol. Mag.*, Decade 5, Vol. 4, pp. 490–500.

Thorarinsson, S. 1953. The crater groups in Iceland. *Bull. Volcanol.*, Série 2, Vol. 14, pp. 1–44.

Tomes, R. F. 1899. Observations on some British Cretaceous Madreporaria, with the description of two new species. *Geol. Mag.*, Decade 4, Vol. 6, pp. 298–307.

Tomkeieff, S. I. 1939. Zonal olivines and their petrogenetic significance. *Mineral. Mag.*, Vol. 25, pp. 229–251.

— 1940. The basalt lavas of the Giant's Causeway district of Northern Ireland. *Bull. Volcanol.*, Série 2, Vol. 6, pp. 89–143.

— and Patterson, E. M. 1947. The Tertiary igneous rocks in the neighbourhood of Ballintoy, County Antrim. Part 1. The Knocksoghey sill; Part 2. The Carrickarade volcano. *Ir. Nat. J.*, Vol. 9, pp. 89–96, 203–212.

— — 1953. The Tertiary igneous rocks in the neighbourhood of Ballintoy, Co. Antrim. Part 3. The coast near Ballintoy Harbour. *Ir. Nat. J.*, Vol. 11, pp. 35–45.

Versey, H. C. 1958. Derived ammonites in basal Cretaceous conglomerate. *Geol. Mag.*, Vol. 95, p. 440.

Visscher, H. 1966. Palaeobotany of the Mesophytic. 3. Plant microfossils from the Upper Bunter of Hengelo, The Netherlands. *Acta Bot. Neerl.*, Vol. 15, pp. 316–375.

— and Commissaris, A. L. T. M. 1968. Middle Triassic pollen and spores from the Lower Muschelkalk of Winterswijk (The Netherlands). *Pollen & Spores*, Vol. 10, pp. 161–171.

Walker, G. P. L. 1959. Some observations on the Antrim basalts and associated dolerite intrusions. *Proc. Geol. Assoc.*, Vol. 70, pp. 179–205.

— 1960. The amygdale minerals in the Tertiary lavas of Ireland. 3: Regional distribution. *Mineral. Mag.*, Vol. 32, pp. 503–527.

— 1962. A note on occurrences of tree remains within the Antrim Basalts. *Proc. Geol. Assoc.*, Vol. 73, pp. 1–7.

— 1971. Compound and simple lava flows and flood basalts. *Bull. Volcanol.*, Vol. 35, pp. 579–590.

WARRINGTON, G. 1970. Palynology of the Trias of the Langford Lodge Borehole. Pp. 44–52 *in* MANNING, P. I., ROBBIE, J. A. and WILSON, H. E. 1970. Geology of Belfast and the Lagan Valley. *Mem. Geol. Surv. North. Irel.*

— 1974. Studies in the palynological biostratigraphy of the British Trias. 1. Reference sections in west Lancashire and north Somerset. *Rev. Palaeobot. Palynol.*, Vol. 17, pp. 133–147.

— 1975. Pp. 123, 125 in *Annu. Rep. Inst. Geol. Sci. for 1974.*

WATTS, W. A. 1962. Early Tertiary pollen deposits in Ireland. *Nature*, Vol. 193, p. 600.

WELCH, R. and WRIGHT, J. 1902. Some Cretaceous foraminifers from north Antrim. *Ir. Nat.*, Vol. 11, pp. 178–180.

WENTWORTH, C. K. and MACDONALD, G. A. 1953. Structures and forms of basaltic rocks in Hawaii. *Bull. U.S. Geol. Surv.*, No. 994. 98 pp.

WHELAN, C. B. 1935. Studies in the significance of the Irish Stone Age; the Campagnian question. *Proc. R. Ir. Acad.*, Vol. 42, Section C, pp. 121–143.

WHITEHURST, J. 1786. *On the original state and formation of the Earth.* 2nd edit. (London.)

WILSON, A. 1902. Notes on blown sand at White Park. *Ir. Nat.*, Vol. 11, p. 196.

WILSON, H. E. 1972. *The regional geology of Northern Ireland.* 113 pp. (Belfast: HMSO).

— and ROBBIE, J. A. 1966. Geology of the country around Ballycastle. *Mem. Geol. Surv. North. Irel.*

WILSON, R. L. 1959. Remanent magnetism of late Secondary and early Tertiary British rocks. *Philos. Mag.*, Vol. 4, pp. 750–755.

— 1961. Palaeomagnetism in Northern Ireland. *Geophys. J.*, Vol. 5, pp. 45–69.

WOLFE, M. J. 1968. Lithification of a carbonate mud; Senonian Chalk in Northern Ireland. *Sediment. Geol.*, Vol. 2, pp. 263–290.

WOOD, C. J. 1967. Some new observations on the Maestrichtian Stage in the British Isles. *Bull. Geol. Surv. G.B.*, No. 27, pp. 271–288.

— 1972. Cretaceous. Pp. 51–58 in *Regional geology of Northern Ireland*, by H. E. WILSON. 113 pp. (Belfast: HMSO).

WOODWARD, S. 1833. *An outline of the geology of Norfolk.* (Norwich.)

WRIGHT, A. M. 1823. *A guide to the Giant's Causeway and the north-east coast of Antrim.* (London.)

WRIGHT, T. 1864–1882. A monograph on the British fossil Echinodermata from the Cretaceous formations. *Monogr. Palaeontogr. Soc.*, Vols. 18, 22, 24–29, 35 and 36.

— 1883. Monograph on the Lias ammonities of the British Isles. Part VI. *Monogr. Palaeontogr. Soc.*, Vol. 37, pp. 401–440.

WRIGHT, W. B. 1919. An analysis of the Palaeozoic floor of north-east Ireland. *Sci. Proc. R. Dublin. Soc.*, Vol. 15, pp. 629–650.

APPENDIX 3

Records of boreholes

CORBALLY RESERVOIR BOREHOLE

Site Approximately 30 m east of the north-east corner of the reservoir [C 879 385]. 6-in map Antrim 6 NE. Height above OD about 70 m. Date 1936. For Portrush U.D.C. Diameter 16 in (40.6 cm) to 152.40 m, 12 in (30.48 cm) to bottom. Information from the late J. J. Hartley and the Town Clerk, Portrush.

	Thickness m	*Depth* m
Basalt, locally red-stained and frequently amygdaloidal	106.68	106.68
Soft green pyritous material, like tuff	12.19	118.87
Lignite	0.23	119.10
Basalt	33.30	152.40
Chalk, recrystallised	2.44	154.84
Basalt	9.14	163.98
Chalk, recrystallised	10.67	174.65
Basalt	9.14	183.79
Chalk, recrystallised	3.66	187.45
Basalt	13.72	201.17
Chalk	24.38	225.55
Basalt	7.62	233.17
Chalk	0.15	233.32
Basalt	14.18	247.50
Chalk	14.63	262.13
Chalk, glauconitic	1.22	263.35
Limestone (sic)	0.61	263.96
Sandstone, pebbly	1.52	265.48
Lias shales, horizontal, pyritous. Slickensided bedding planes	10.36	275.84

Note No record is preserved of the water levels or yields, but the well was never used and the yield must have been very small.

PORT MORE BOREHOLE

Site Coolmaghra Townland, 100 m south-south-west of Glenstaghey Lower Farm [D 069 435]. 6-in map Antrim 4 SE. Height above OD, 103 m. Date, 1965–67. For Government of Northern Ireland.

	Thickness m	*Depth* m
Soil and Glacial Drift		
Alluvial clay	1.22	1.22
TERTIARY		
Lower Basalts		
Basalt (drilled by rock-bit)	19.81	21.03
Basalt, with abundant zeolites	2.44	23.47
Basalt, compact, scattered zeolites	10.44	33.91
Tuff, lateritised at top, with fragments of Lias, Chalk, basalt and laterite. Banded.	15.90	49.81
Basalt, probably two flows	20.70	70.51
Tuff, basaltic, with cobbles of fresh basalt	6.12	76.63
Clay-with-Flints		
Flint nodules up to 12 cm in grey clayey matrix	0.58	77.21
CRETACEOUS	*Thickness* m	*Depth* m
Ulster White Limestone Formation (Chalk)		
Chalk, white, intensely hard, with courses of flint and irregular stylolitic and wavy partings (post-Larry Bane Chalk sequence). Glauconitised hardground (North Antrim Hardground) at 114.68 m	56.06	133.27
Chalk, as above, divided into two units, the upper twice the thickness of the lower, delimited by well-marked erosional planes (Larry Bane Chalk)	7.62	140.89
Chalk with stylolitic planes and courses of flints as above, with many tabulate flints in lower part (pre-Larry Bane Chalk sequence). Glauconitised pebble bed (Bendoo Pebble Bed) at 162.76 m; below this depth chalk, grey, gritty with comminuted *Inoceramus* shells, with glauconite content increasing downward. Phosphatised pebbles, pyrite masses and algal stromatolitic structures in basal 15 cm (Basal Conglomerate)	27.46	168.35
JURASSIC		
Lower Lias		
Mudstone, dark grey, fossiliferous, slightly calcareous. Ironstone nodules	2.49	170.84
Shale, grey-black, micaceous	3.40	174.24
Mudstone, buff with black, pyritic burrow-traces	6.58	180.82
Fault Breccia, 45° inclination	0.08	180.90
Mudstone, shaly, dark grey with thin beds of buff siltstone. Ironstone nodules; veins of pyrite; some listric surfaces and calcite veins	40.67	221.56
Mudstone, grey, often slightly calcareous. Thin limestone and calcareous siltstone ribs common	75.82	297.38
Mudstone, grey. Rare calcareous ribs	40.19	337.57
Dyke		
Basalt, steep-dipping, pyrite vein	0.15	337.72
Mudstone, grey, calcite veined	3.17	340.89
Sill		
Basalt, fine-grained. Zeolite and calcite veins	4.27	345.16
Dolerite, coarse-grained and locally zeolitic	16.13	361.29
Mudstone, hornfelsed	0.10	361.39
Mudstone, grey, often calcareous, with a few thin limestone bands in upper part. Ironstone nodules and sparse pyrite	75.90	437.29
Mudstone, hornfelsed	0.51	437.80
Sill		
Basalt with xenoliths of hornfels	1.47	439.27
Dolerite	1.52	440.79
Basalt in vertical contact with hornfels	1.09	441.88
Basalt and dolerite	1.60	443.48
Basalt in vertical contact with hornfels	1.02	444.50
Basalt with xenoliths of hornfels and veins of hornfels and fine-grained basalt	15.24	459.74
Basalt, sometimes pyritous, slickensided, with vein calcite on joints	7.82	467.56
Dolerite, often decomposed to spheroidal gravel and sand	27.43	494.99
Basalt, fine and coarse-grained, possibly banded	9.23	504.22
Basalt with abundant inclusions of greenish spotted siltstone, and mudstone	12.77	516.99

	Thickness m	Depth m
Basalt, fresh	14.78	531.77
Basalt with hornfels rafts	0.74	532.51
Hornfels	1.98	534.49
Basalt with a few hornfels rafts	9.35	543.94
Basalt with a large proportion of hornfels	5.97	549.91
Dolerite, variable in size, commonly decomposed to sand	39.52	589.43
Hybrid rock of melanocratic basalt with blocks of leucocratic material	1.07	590.50
Dolerite, largely decomposed, with much zeolitic material. Some fresher bands and rare wisps of possible hornfels	65.35	655.85
Basalt, broken and with hornfels rafts in lowest 60 cm	4.85	660.70

TRIASSIC

Lower Rhaetian

WESTBURY FORMATION

	Thickness m	Depth m
Mudstone, shaly, dark grey with grey-green siltstone partings. *Rhaetavicula contorta* abundant in mudstones. A little iron pyrite	2.75	663.45
Mudstone, shaly, almost black. Pyrite coatings on bedding-planes	1.32	664.77

Mercia Mudstone Group

COLLIN GLEN FORMATION

	Thickness m	Depth m
Marl, massive, compact, greenish-grey passing in lowest 30 cm into red-brown. *Euestheria sp.* aff. *portlocki*	5.69	670.46
Marl, alternations of green-grey and dull red-brown colour, some mottling. Thin bands of penecontemporaneous breccia and porcellanous-textured cementstones, also locally brecciated, and up to 20 cm thick	5.13	675.59

PORT MORE FORMATION

	Thickness m	Depth m
Marl, dull reddish-brown, local greenish mottling	6.60	682.19
Marl, reddish-brown with calcareous streaks. Rare green reduction spots	15.24	697.43
Marl, silty, red-brown with rare green bands and spots. Pale, wispy, calcareous bands in places. Quartz grains and pebbles common	2.09	699.52
Marl, red-brown, calcareous streaks	3.63	703.15
Quartz pebbles and grains in marl matrix	0.02	703.17
Marl, reddish-brown with occasional green reduction spots and bands; pale calcareous wisps and laminae. Rare quartz grains and mica flakes. Small 'pressure domes' occur rarely on bedding planes	10.49	713.66

KNOCKSOGHEY FORMATION

	Thickness m	Depth m
Marl, red-brown with green reduction spots ('fish eyes') and patches. Silty and micaceous bands. Anhydrite occurs as thin veins and blue-white nodules up to 15 cm in diameter but usually smaller. Thin veins of fibrous gypsum. Tendency for the percentage of anhydrite to decrease downwards, but never more than 5 per cent of total core	64.49	778.15
Marl, reddish-brown with anhydrite nodules and impregnations; thin gypsum veins. Green reduction spots and rare buff-coloured silty patches	34.37	812.52
Siltstone, pale green; red-brown marl with sand grains; and red-brown non-calcareous mudstone	2.72	815.24
Sandstone, greenish grey, dolomitic. Finely laminated with mudstone bands and flattened pellets (COOLMAGHERA SKERRY)	2.54	817.78
Mudstone, red-brown with green reduction spots. Sparse anhydrite nodules and thin gypsum veins	1.67	819.45
Marl, red-brown, green 'fish eyes'. Thin gypsum veins, rare anhydrite	6.86	826.31
Mudstone, red-brown with green 'fish eyes'.	1.37	827.68
Marl, silty, dull brown with green reduction spots. Dolomitic patches	8.62	826.30

GLENSTAGHEY FORMATION

	Thickness m	Depth m
Mudstone, interbanded, dull or chocolate brown and blue-green. Heavy gypsum veining locally. Thin (5 cm) sandstone bands at 842 m and 843 m show ripple marks and current bedding	14.14	850.44

Fault Rock

	Thickness m	Depth m
Mudstone, silty, brown. Greenish colouration on joints	3.86	854.63
Mudstone breccia. Sharply angular fragments slightly disorientated, up to 7 cm long in silty paler matrix, impregnated with gypsum, 25 cm band of calcareous silty mudstone in middle	1.93	856.56
Mudstone, brown, silty, with some green colour on joints. Scarce anhydrite. Brecciated basal layer	0.74	857.30
Sandstone, pale green and buff	0.18	857.48
Mudstone, silty brown, with some sand grains and small anhydrite nodules	4.62	862.10
Rhythmically banded sandstone–mudstone sequence with five buff sandstone bands up to 20 cm thick interbedded with silty mudstones and siltstones. Disseminated anhydrite and thin gypsum veins	7.93	870.03
Mudstone, dark chocolate-brown, with green intercalations and thin green sandstone bands	5.18	875.21
Siltstone and mudstone in thin alternating bands with rare thin sandstones. Scanty anhydrite	4.67	879.88
Marl, red-brown, in thick beds with thin sandstone and siltstone partings. Small nodules and anhydrite impregnations in upper part. Sparse thin gypsum veins	8.69	888.57
Mudstone, dull and chocolate brown with some blue-green patches and bands. Sparse silty bands and sand grains. A few thin sandstone bands in lowest 7 m	19.00	907.57
Sandstone, pale, banded with silty layers. Small patches of anhydrite (SIXTEEN-INCH SKERRY)	0.40	907.97
Mudstone, dark brown with thin green siltstone and sandstone bands. Scattered anhydrite and gypsum veins	19.36	927.33
Mudstone with silty layers and a 5-cm limestone band, showing increasing signs of thermal metamorphism downwards, the lowest 23 cm being completely hornfelsed	4.85	932.18

Sill

	Thickness m	Depth m
Dolerite, fine-grained at contact with hornfels (inclined at 20° to horizontal), becoming coarser and well jointed below	0.74	932.92

	Thickness m	Depth m
Dolerite, alternating, fresh and decomposed, sometimes altered to a sand. Decomposed bands are up to almost 2 m thick—fresh bands similar. Very poor recovery of last 6 m: drilling suggests that the bottom of the intrusion is very decomposed	27.20	960.12
Hornfels, banded with gypsum veins	0.18	960.30
Mudstone, dark chocolate-brown, becoming redder towards base. Silty and sandy bands and wisps. Signs of metamorphism in colour changes throughout	8.45	968.75
Mudstone, mainly brown but green bands, mottling and reduction spots. Some calcareous nodules and laminae. Some quartz grains in places	10.34	979.09
Mudstone, red-brown and chocolate with silty bands and laminae, apparently dolomitic. Greenish reduction spots. Sandstone bands near base. Some gypsum veins	9.68	988.77
Sandstone (FAREWELL SKERRY) pale with micaceous laminations	0.89	989.66
Mudstone, chocolate, red-brown and grey silty laminae and anhydrite in patches and specks, some of which may pseudomorph halite crystals. Gypsum in thin bands and veins in lower half of succession	20.09	1009.75
Mudstone, grey-green with thin gypsum bands	1.55	1011.30
Mudstone, red-brown with some green and grey bands. Gypsum veins and specks. Halite pseudomorphs near base	16.00	1027.30
Mudstone, grey and rarely, brown. Pale bands of dolomitic siltstone and breccia. Halite pseudomorphs	5.69	1032.99
Mudstone, brown, grey and green, showing signs of thermal alteration. Scattered bluish anhydrite. Missing core makes depth approximate	2.72	1035.71
Sill		
Dolerite, very decomposed and largely missing. 76 cm of compact, medium-grained dolerite with horizontal bottom contact	4.78	1040.49
Mudstone, dark grey, hornfelsed. Anhydrite patches	0.23	1040.72
Mudstone, brown and green with silty and sandy bands. Signs of thermal alteration in uppermost 2.5 m. Specks of anhydrite. Thin band of mudstone breccia at base	8.53	1049.25
Mudstone and siltstone, laminated succession in bands usually 1 to 5 cm thick. Desiccation cracks. Occasional anhydrite patches	1.27	1050.52
Mudstone, red-brown, rarely green	5.61	1056.13
CRAIGANEE FORMATION		
Siltstone and sandstone, pale green. Anhydrite nodules (THREE-FOOT SKERRY)	1.07	1057.20
Mudstone, red-brown, with paler silty bands and green reduction spots	5.94	1063.14
Sandstone, pale grey-green with blue anhydrite patches	0.46	1063.60
Mudstone, red-brown with rare green laminae, with silty and sandy laminations. Scattered anhydrite patches	5.64	1069.24
Sandstone, pale grey with blue-green streaks	1.27	1070.51
Mudstone, red-brown with thin greenish bands with siltstone bands giving striped appearance. Rare thin sandstones	5.94	1076.45
Sandstone, pale with green laminae. Some marl partings and anhydrite patches	1.42	1077.87
Mudstone, red and chocolate-brown. Some green reduction spots. Thin bands of flow breccia and sandstones up to 30 cm thick	18.37	1096.24
Sandstone, pale, with brown siltstone partings. Nodules of anhydrite	1.02	1097.26
Mudstone, brown with siltstone partings and sandy bands	2.06	1099.32
Sandstone, pale, with mudstone partings	0.86	1100.18
Mudstone, red-brown, with siltstone partings	2.99	1103.17
Sandstone, pale green	0.44	1103.61
Siltstone, brown	0.48	1104.09
Sandstone and siltstone partings	0.89	1104.98
Mudstone, red-brown, sometimes laminated. Rare salt pseudomorphs. Siltstone bands	16.12	1121.10
Sandstone, pale with red and green partings	1.40	1122.50
Mudstone, red-brown, and sandstone bands	1.98	1124.48
Sandstone, pale with mudstone partings	0.36	1124.84
Mudstone, chocolate brown. Silty and sandy partings	0.61	1125.45
Sandstone, brown, with dark mudstone laminae	1.14	1126.59
Sandstone, buff (SUITCASE SANDSTONE)	0.91	1127.50
Mudstone, dark, with paler siltstone and buff sandstone bands	1.45	1128.95
Mudstone, red to chocolate-brown, with siltstone and sandstone bands	9.68	1138.63
Sandstone, buff and brown with mudstone bands	2.31	1140.94
Mudstone, red-brown, with green silty partings and green mottling. Ripple marks and mudcracks	4.60	1145.54
Mudstone, siltstone and sandstone layers; usually a few centimetres thick, with sandstone units up to 1.5 m thick at about 2-m intervals. Argillaceous beds are usually red-brown: sandstones buff or off-white. Anhydrite nodules occur in the sandstones. Mudstone breccias and mudcracks in the argillaceous beds	77.36	1222.9
LAGAVARRA FORMATION		
Sandstones, mudstones, siltstones thin-bedded with slump-structures, auto-brecciation, salt pseudomorphs. Colour mainly red-brown, occasionally green. Small anhydrite nodules and specks common. Individual beds rarely over 50 cm	94.42	1317.32

Sherwood Sandstone Group

UPPER SANDSTONE FORMATION

	Thickness m	Depth m
Sandstones, red-brown or buff. Mudstone rafts, current-bedding. Quartz pebbles common towards base. Mudstone partings.	7.03	1324.35
Conglomerate, polygenetic, with pebbles up to 10 cm of metamorphic rocks, rhyolite, quartz and mudstone. Pebbly sandstone bands and mudstone partings	2.03	1326.38
Sandstone, brown and occasionally pebbly with siltstone and mudstone partings	4.88	1331.26
Conglomerate, polygenetic but mainly of quartz and quartzite cobbles and pebbles. Cobbles up to 10 cm, but largely pebble-grade in sandy matrix. The conglomerate is piped 13 cm into the top of the underlying sandstone	3.48	1334.74

	Thickness m	Depth m
Sandstone, red-brown, laminated, pebbly	0.91	1335.65
Breccia. Sandstone fragments and gypsum veins	0.59	1336.24
Fault		
Sandstone, brown and white with pebbles and mudstone wisps. Bedding dips steeply at 45°	2.29	1338.53
Sandstone, pale brown, buff and off-white. Mudstone clasts and wisps occur at many levels. Sporadic pebbles, usually quartz and quartzite. Bedding and lamination usually horizontal; but some false bedding seen. Thin mudstone and siltstone partings towards the base	21.34	1359.87
Sandstones, red-brown with abundant mudstone clasts and sporadic pebbles. Thin mudstone 'wayboards' usually some 5 cm thick often showing sand-filled desiccation cracks. Sandstones commonly show false bedding and erosional contacts	23.19	1383.06
Sandstone, coarse, with granite pebbles up to 2½ cm	0.25	1383.31
Sandstone, brown and pink. Sporadic pebbles and mudstone clasts. Cross-bedding and rare thin mudstone 'wayboards'	6.68	1389.99
Sandstone, mainly red-brown but darker and paler bands. Mudstone clasts and rare wayboards. Rare pebbles. A few bands emit foetid odour when broken	31.90	1421.89
Sandstone, purple and grey at top, passing down into normal red-brown. Mudstone clasts and rare quartz pebbles	15.95	1437.84
Sandstone, pale brown and pink. Notably coarse and pebbly. Cross-bedding. Chocolate-brown mudstone clasts	7.80	1445.64
Mudstone, chocolate-brown, micaceous	0.28	1445.92
Sandstone, purple-brown to pink. Mudstone clasts, rare pebbles. Autobreccia of brown medium-grained sandstone fragments in finer sandy matrix with small pebbles	7.47	1453.39
Sandstones, pale to dark brown. Some pebbles. Thin mudstone wayboards and clasts. Cross-bedding. Rare siltstone bands	24.31	1477.70
Sandstone, massive, mainly pale brown. Mudstone clasts and small pebbles ubiquitous	24.03	1501.73
Siltstone and fine sandstone. Small anhydrite nodules	3.99	1505.72
Dyke?		
Basalt, fine-grained with fragments of hornfels. Recovery poor.	2.51	1508.23
Sandstone, pebbly, pale, with mudstone partings and clasts. Signs of baking throughout	3.53	1511.76
Sandstone, pale, coarse and pebbly. Thin mudstone layers	2.54	1514.30
Sandstone, red-brown, fine-grained with fine mudstone wisps and clasts. Sparse pebbles. Current bedding	64.84	1579.14
Fault plane dipping at 45°		
Sandstone, dark but with pale bands. Mudstone wayboards, sparse pebbles	6.25	1585.39
CONGLOMERATE FORMATION		
Conglomerate, pale brown. Sub-rounded pebbles of quartz, quartzite, epidiorite and rarely felsite and andesite, in a sandy matrix. Partings of red-brown pebbly sandstone up to 30 cm thick and a few mudstone bands, usually a few cm but rarely up to 15 cm	27.89	1613.28
Conglomerate, dark brown. Rounded pebbles of andesite and quartzite with interstitial anhydrite and dolomite. Sandstone partings, usually less than 10 cm thick	35.94	1649.22
Dyke		
Basalt, fine-grained, fresh, steeply inclined upper junction with conglomerate	5.31	1654.53
Conglomerate, dull-brown. Pebbles of andesite and quartzite. Thin partings of pebbly grit and one of mudstone. Dolomitised	8.79	1663.32
LOWER SANDSTONE FORMATION		
Sandstone, red-brown, grey and buff coloured beds. Cross-bedding. Slight foetid odour when freshly broken	13.84	1677.16
Sandstone, red-brown, sparsely pebbly. Thin mudstone partings and some bands of sandstone with mudstone clasts	36.20	1713.36
Mudstone, red-brown, silty and often auto-brecciated. More or less dolomitic throughout	11.96	1725.32
Sandstone, red-brown, with rare chocolate mudstone bands and scattered pebbles. Dolomitic cement in upper part, often in pseudobrecciated beds	43.21	1768.53
Conglomerate and sandstone in thin beds, rarely over 60 cm thick. Sandstone and matrix red-brown, medium-grained. Pebbles of schist, quartz and quartzite, rarely dolomitic limestone. The proportion of conglomerate falls towards the bottom of the section and a few thin mudstone and siltstone wayboards appear	61.79	1830.32

?PERMIAN

MARLS

	Thickness m	Depth m
Mudstone, dark red-brown, silty and micaceous. Dolomitic patches and thin ribs of dolomitic limestone near top	14.76	1845.08
SANDSTONE		
Sandstone, pebbly	0.43	1845.51
Dyke		
Basalt, fine-grained, decomposed	0.31	1845.82
Sandstone, conglomeratic, with pebbles of quartz, quartzite, pink dolomite and schist, in thin beds. A few thin mudstone bands near bottom	19.30	1865.12
Striped measures—thin beds of purple-brown mudstone, red-brown siltstone, and red-brown sandstone, rarely pebbly. Beds rarely over 10 cm thick, usually less	2.44	1867.56
Sandstone, often conglomeratic, with schist, quartz and quartzite pebbles. Rare thin mudstone partings in upper part. A high proportion of grit-grade sandstone grains—over 1 mm—often well rounded *Seen to*	29.13	1896.69

APPENDIX 4

Chemical analyses of rocks

	1	2	3	4	5	6	7	8	9	10	11	12
SiO_2	44.87	52.1	45.93	45.34	46.57	20.60	9.78	10.15	8.82	21.82		30.04
Al_2O_3	14.88	14.7	15.22	14.67	15.60	45.50	17.82	20.11	31.41	27.13		26.57
Fe_2O_3	2.23	3.2	3.06	2.40	0.99	4.61	47.32	46.02	23.39	27.81		18.95
FeO	10.57	7.4	9.26	9.15	9.14	n.d.	0.62	0.90	0.50	0.77		1.25
MgO	11.51	6.5	10.30	13.32	9.05	tr.	0.33	0.75	0.34	0.41		0.60
CaO	9.79	10.8	8.55	9.12	11.77	tr.	1.23	0.64	0.65	0.29		0.26
Na_2O	2.20	2.8	2.63	1.86	1.89	—	—	—	—	—		—
K_2O	0.12	0.8	0.19	0.24	0.72	—	—	—	—	—		—
TiO_2	1.39	1.0	1.54	1.13	0.84	5.18	5.12	4.58	2.05	2.53		2.10
P_2O_5	0.23	0.5	0.28	0.09	0.52	—	—	—	—	—		—
MnO	0.20	0.2	0.18	0.22	0.16	0.05	0.06	0.04	0.05	0.08		0.08
H_2O+	1.80	—	1.68	1.69	1.79	23.60	8.93	9.75	16.97	13.72		10.55
H_2O-	1.11	—	0.89	1.05	0.75	—	7.86	6.03	13.64	5.37		8.84
	100.90	100.0	99.71	100.28	99.79	99.54	99.07	98.97	97.82	99.93		99.24
Kaolin-type mineral						44.3	5	—	—	10	—	—
Halloysite						—	—	—	—	—	70	50
Illite						—	10	15	10	10	—	—
Montmorillonite						—	5	5	10	—	10	5
Gibbsite						42.9	10	10	35	30	5	5
Hematite						—	45	40	15	30	5	15
Goethite						—	—	—	5	—	—	—

1 Lower Basalts, Ballykeel (Patterson, 1951)
2 Tholeiitic Basalts, average for seven flows, calculated water-free (Patterson, 1955)
3 Upper Basalts, Craig Park (Patterson, 1951)
4 Upper Basalts, Dunmull (Patterson, 1951)
5 Portrush Sill (Holmes, 1936)
6 Siliceous bauxite, Clegnagh (Eyles, 1952)
7 Lateritic iron ore and ferruginous bauxite. Ballylagan; top of bed (Patterson, 1955)
8 As No. 7; 38 cm below top of bed (Patterson, 1955)
9 As No. 7; 76 cm below top of bed (Patterson, 1955)
10 Bauxite lithomarge and lithomarge, Upper Interbasaltic Bed, Croaghmore; top (Patterson, 1955)
11 As No. 10; middle (Patterson, 1955)
12 As No. 10; base (Patterson, 1955)

APPENDIX 5

Geological survey photographs

Copies of these photographs, which were taken mainly by J. M. Pulsford with a few by H. E. Wilson, are deposited for public reference in the office of the Geological Survey, 20 College Gardens, Belfast, and in the Libraries of the Institute of Geological Sciences at Exhibition Road, London SW7, and in Leeds and Edinburgh. Prints and colour transparencies can be supplied.

NI 242 Basalt lavas on the foreshore, Portstewart.
NI 243 Vent agglomerate. The Berrins, Portstewart.
NI 244 Lower Basalt scarps on foreshore. Black Rock.
NI 245 Volcanic vent punched through bedded lavas. Devil's Port.
NI 246 Vent agglomerate on foreshore. Craigtown Beg.
NI 247 Block of recrystallised limestone in weathered top of lava flow. Craigtown Beg.
NI 248 Volcanic vent and dyke in bedded lava. Craigtown Beg.
NI 249 Columnar Tholeiitic Basalt lava. Crossreagh Quarry.
NI 250 Panoramic view of the high ground. Carnalridge, Portrush.
NI 251 Estuarine clay with shells. Bann estuary.
NI 252 Estuarine sands and blown sand dunes. Bann estuary.
NI 253 Drift above striated pavement at top of basalt lavas. Carnanee Quarry.
NI 254 Dune sands and beach. Portstewart.
NI 255 Bann Estuary. Ballycavin.
NI 256 Entrance to iron-ore workings in the Interbasaltic Bed. Ballylagan Mine.
NI 257 Residual basalt spheroid in the Upper Interbasaltic Bed. Ballylagan Mine.
NI 258 Portrush Sill. Ramore Head.
NI 259 Vertical contact of Portrush Sill with indurated Lias shale. Reviggerly.
NI 260 Jointing in Portrush Sill. Reviggerly.
NI 261 Indurated Lias shale overlying massive dolerite sill. Kerr St. Quarry, Portrush.
NI 262 Portrush Rock: indurated Lias shales. Blue Pool.
NI 263 Ammonites in indurated Lias shale. Blue Pool.
NI 264 Raft of contorted indurated Lias shale in top of dolerite sill. Castle Island.
NI 265 Segregation sheet near top of dolerite sill. Large Skerries.
NI 266 Horizontal layering in dolerite sill. Large Skerries.
NI 267 Rounded blocks of basalt embedded in shattered Chalk. Stradly Rock.
NI 268 Agglomerate vents passing through shattered Chalk. Stradly Rock.
NI 269 Irregular top surface of the Chalk. Long Gilbert Quarry.
NI 270 Irregular intrusions of basalt in Chalk. Long Gilbert Quarry.
NI 271 Hollow in chalk surface infilled with basalt rubble. Long Gilbert Quarry.
NI 272 Sheelas Head. Chalk cliffs and caves. Portnool.
NI 273 Chalk cliffs and caves and marine rock-platform. Sliddery Cove.
NI 274 Marine erosion in Chalk. Sliddery Cove.
NI 275 Volcanic agglomerate. The Pound, Dunluce Castle.
NI 276 Banded lava flow. The Sugarloaf, Dunluce.
NI 277 Agglomerate vent cutting through banded lava. Gortnabane Cove.
NI 278 Dyke 6 m wide forming a stack on foreshore. Portnaboe.
NI 279 Spheroidal weathering in olivine-basalt. Lower Basalt Series, Weir's Snout.
NI 280 Lower Basalt Lavas. Great Stookan.
NI 281 Relict basalt spheroid in lower part of Interbasaltic Bed. The Highlandman's Bonnet.
NI 282 The Giant's Causeway. Windy Gap.
NI 283 Polygonal jointing in Tholeiitic Basalt lava. Giant's Causeway.
NI 284 Columnar basalt forming the Middle Causeway.
NI 285 Columnar tholeiitic basalt lava. Giant's Loom.
NI 286 Base of the Giant's Causeway flow of tholeiitic basalt.
NI 287 Columnar tholeiitic basalt. Giant's Organ.
NI 288 Roveran Valley Head and Spaniard Rock. Giant's Causeway.
NI 289 Roveran Valley Head: dolerite dyke.
NI 290 Spaniard Rock and the Chimney Tops. Port Reostan.
NI 291 Columnar Tholeiitic Basalt Lavas. The Amphitheatre.
NI 292 Curved columns of tholeiitic basalt. Irish Harp and Chimney Tops.
NI 293 Spheroidal weathering of residual basalt cores in Interbasaltic Beds. 'Giant's Eyes'.
NI 294 Giant's Causeway with Roveran Valley Head. Spaniard Rock from Weir's Snout.
NI 295 Giant's Causeway from above.
NI 296 Benanouran Head with Pleaskin Head and Benbane Head.
NI 297 Pleaskin Head with Benbane Head in background.
NI 298 Pleaskin Head.
NI 299 Basalt coastal scenery near Bengore Head.
NI 300 Lower flows of Tholeiitic Basalts. Portmoon Bay.
NI 301 Basalt coastal scenery at Portmoon Bay.
NI 302 Details of surface weathering in a 'ropy' lava. Dunseverick.
NI 303 Raised-beach platform in Tholeiitic Basalts. Dunseverick.
NI 304 Columnar jointing at Ballynastraid Quarry. Whitepark Bay.
NI 305 Intrusive sheet in Lower Basalt. Portballintrae.
NI 306 Boulder clay and glacial gravels. Portballintrae Bay.
NI 307 Details of a Silt-ball in boulder clay. Portballintrae.
NI 308 Blown sand dunes extending across mouth of River Bush.
NI 309 Glacial spillway near Causeway Head.
NI 310 Exposure of Interbasaltic Bed near Ballaght.
NI 311 White Park Bay.
NI 312 Upper Chalk and Basalt coastal scenery. White Park Bay.
NI 313 Sea stacks in Tholeiitic Basalts. Ballintoy Harbour.
NI 314 Detail of volcanic ash and lava succession at Ballintoy.
NI 315 Upper Chalk near Ballintoy.
NI 316 Chalk/Basalt coastal scenery at Ballintoy Harbour.
NI 317 Upper Chalk and Tertiary dolerite intrusions. Ballintoy.
NI 318 Chalk and Tholeiitic Basalts faulted together. Ballintoy Harbour.
NI 319 Chalk and Basalt scenery near Ballintoy Harbour.
NI 320 The Knocksoghey Sill.
NI 321 Western margin of the Carrickarade volcanic vent.
NI 322 General view of Carrickarade volcanic vent.
NI 323 Carrickarade Tertiary volcanic vent.

NI 324	Volcanic agglomerate with associated intrusive basalt. Carrickarade.
NI 325	'Submerged Forest'. Mill Strand, Portrush.
NI 326	Submerged Peat. Mill Strand, Portrush.
NI 327	Well bedded iron-stained blown sand overlying 1.2 m of peat. Portrush.
NI 328	Sand dunes with the Skerries in the distance.
NI 329	Lignite at top of Interbasaltic Bed. Craigahullier Quarry.
NI 330	Old Quarry. Craigahullier.
NI 331	Columnar lava. Craigahullier.
NI 332	Working face in poorly columnar lava. Craigahullier.
NI 333	Stratified glacial gravels in disused pit near Bushmills.
NI 334	Junction of Middle and Upper Basalts near Bushmills.
NI 335	Flow 4 of the Tholeiitic Basalts near Bushmills.
NI 336	Craig Hill Quarry in a massive flow of the Upper Basalts.
NI 337	Crushing, screening and coating plant. Craig Hill Quarry.
NI 338	Upper Interbasaltic Bed. Croaghmore Hill.
NI 339	Upper Basalt Quarry near White Park Bay.
NI 436	Spotted dolerite with irregular dark olivine-rich clots. Large Skerries.
NI 437	Portrush Sill. Large Skerries.
NI 438	Portrush Sill. Castle Island.
NI 439	Composite vein or sheet about 60 cm wide in massive dolerite.
NI 440	Composite vein or sheet in massive dolerite.
NI 441	Liassic shales, now contorted hornfels. Castle Island.
NI 442	Banded dolerite at the base of the middle 'granophyric' zone of sill. Little Skerries.
NI 443	Scarp feature formed by upper [massive dolerite] zone of sill. Little Skerries.
NI 444	Composite vein 7.6 cm wide in massive dolerite. Little Skerries.
NI 445	Little Skerries, showing scarp features formed by massive layers in sill.
NI 446	Spherical area of pale weathered dolerite in massive well jointed dolerite. Ramore Head.
NI 447	Rough-weathering dolerites from lowest exposed part of sill. Ramore Head.
NI 448	Portrush Sill. Lower part of sill, showing faint banding. Ramore Head.
NI 703	Basalt dyke forming raised-beach stack. Portnaboe.
NI 704	Lower Basalt lavas and Interbasaltic Bed. Port Ganny.
NI 705	The Grand Causeway; view from Middle Causeway.
NI 706	Columnar tholeiitic basalt cross-joints. Giant's Causeway.
NI 707	Top of Grand Causeway. Giant's Causeway.
NI 708	Columnar jointing in tholeiitic basalt. Giant's Causeway.
NI 709	Spheroidal weathering of olivine-basalt. Windy Gap.
NI 710	Middle Causeway—columnar tholeiitic basalt lava.
NI 711	Spaniard Rock. Lower Basalts.
NI 712	Footpath cut in Interbasaltic Bed. The Amphitheatre.
NI 713	Columnar 'colonnade'. Tholeiitic Basalts. The Organ.
NI 714	Columnar basalt. Giant's Causeway.
NI 715	Spheroidal weathering. Windy Gap.

INDEX

Aa lava 23–24, 29
Aeolian sandstones 6
Aeromagnetic survey 145
Agglomerate 23, 31, 38, 131, 133
Ailsacraig 121, 126
Aird 124
Aird Snout 55, 124
Alluvium 3, 46, 49, 50
Altachuile Bay 90, 99
Altachuile Breccia 16, 86, 90, 96–97
Amphitheatre 118, 124
Anhydrite 6
Antrim Coast Readvance 3, 43
Antrim Lava Series 23–29, 116–132, 150
Ardina 44, 46, 127
Ardmore Anticline 143
Armoy Bog 45, 50
Armoy–Ballymoney moraine 43, 45
Articlave River 44, 46, 49, 56, 68, 127

Ball lava 116
Ballintoy 4, 5, 11–14, 16, 19, 23–26, 31, 33, 40, 42, 47, 51, 55, 58–60, 78, 80, 85, 88, 90, 97, 98, 118–124, 135
Ballintoy Chalk 15, 16, 85, 96–98
Ballintoy Harbour Fault 25, 26, 42, 119
Ballyaghran Point 47, 48
Ballyallaght 121
Ballycastle 6, 11, 17, 42, 54, 143
Ballycastle Chalk 111
Ballycastle–Ballymoney Fault 29
Ballycraig 24, 26, 60, 117, 121, 125
Ballygallin 127
Ballyhome 27, 125, 129
Ballykeel 117, 121
Ballylagan 29, 60, 68, 125, 126, 129, 132
Ballyleese 127
Ballylough 45
Ballymaclevennon 127
Ballymacrea 39, 117
Ballymagarry Chalk 15, 17, 87, 109–110
Ballymagarry Quarry 17, 38, 51, 68, 87, 90, 99, 100, 108–111, 116
Ballynastraid 31, 118, 121, 126
Ballyoglagh 130
Ballyreagh 29, 127, 129
Ballytober 126
Ballyversall 44
Ballywatt 44
Ballywillan 50, 117, 119, 125
Bann Brook 46, 48, 49
Bann River 2, 3, 27, 42–45, 56, 58, 127
Barrows 58
Basal Conglomerate 15, 16, 68, 90
Basalt 23–29, 31, 44, 51
Basin-bog 50
Bauxite 23, 26, 51, 60, 68, 119–124, 126, 132
Beardiville 44, 58
Beeston Chalk 17, 99, 100
Belfast 15, 48, 85

Benadir 125
Benanouran Head 68, 123–24
Benbane Head 24, 26, 42, 55, 118, 124
Bendoo 16, 19, 24, 33, 38, 47, 92, 93, 95, 139
Bendoo Pebble Bed 16, 17, 85, 88, 92, 93, 112
Bengore Head 5, 10, 20, 125
Berginan's Port 116
Berrins, The 127, 134
Binard 31, 133
Black Hill Port 129
Black Knowe 29, 126
Blackpool 77
Black Park 126
Black Rock 40, 128, 137, 142
Blown sand 3, 48
Blue Pool 11, 39
Boehmite 132
Boheeshane Bay 13, 16, 33, 78, 85, 88, 90–94
Boheeshane Chalk 15–17, 85, 86, 88, 93–96
Bole 23
Bonagarry 116
Boneyclassagh 117, 121, 126
Bouguer anomalies 143
Boulder Clay 44, 56
Brick Clay 44
Brocks Cove 135
Bronze Age 47, 48, 58
Brookvale Terrace 48
Bruce, Robert the 59
'Bunter' 2, 6
Burnfoot 116
Burrow-fills 91, 92, 93, 98
Bush, River 2, 43, 46, 47, 49, 50, 57, 58, 130
Bushfoot 16, 24, 55, 58, 138
Bushmills 44, 45, 48, 58, 59, 85, 121, 126
Bushmills Electric Tramway 59

Cairns 58
Calcite 24, 68
Caledonian orogeny 2
Cappagh 127
Carboniferous 43, 44, 47, 156
Carnaboy 44, 45
Carmean 100
Carnalbanagh 43
Carnanee 29, 44, 45, 47, 48, 57
Carrickarade 2, 5, 17, 23, 24, 26, 29–31, 33, 34, 38, 54, 55, 59, 79, 118, 119, 123, 131, 133, 139
Carrowcloghan 44
Carrowreagh 24, 45, 124, 130
Cashels 58
Castle Erin 47, 48
Castle Island 39, 137, 138, 139, 140
Catton Sponge Bed 17, 98
Causeway Head 123, 135
Caves 47
Celadonite 68
Cenomanian 3, 15
Chalcedony 68, 130
Chalk 3, 15–19, 44, 51, 68, 85–115, 154
Chimney Tops 124

Church Bay 94, 95
Clay-with-flints 23
Clegnagh 51, 60, 68, 121–122
Cloghastucan Chalk 15, 16, 85–88, 90, 91–92
Clooney 117
Cloughorr 117
Cloyfin 28, 44, 45, 50, 60, 126
Coccoliths 15, 92
Collin Glen Formation 6, 74–76
Colonnade 27, 28, 124, 126
Columnar basalt 4, 27, 28
Compound flows 24, 116
Conglomerate Formation 6, 7
Contham Head 118
Coolmaghera 123
Coolmaghra Skerry 6, 75, 154
Cooraghy 97
Corbally 11, 26, 39, 51, 117, 119, 120, 129, 140
Corrstown 11
Court grave 58
Craig Hill 130
Craigahullier 4, 26, 39, 51, 60, 68, 117, 119–121, 125, 131
Craigalappen 126
Craiganariff 29, 127
Craiganee Formation 6, 7, 74, 76, 154
Craignahorn 127
Craigtown 38, 44, 127, 134, 138
Crannogs 58
Creggan Chalk 15–17, 85–88, 90, 92–93
Crescent, The 47, 56, 128
Cretaceous 3, 5, 15–19, 84–115
Croaghmore 29, 126, 130
Crocknamolt 29, 60, 121, 125
Crossreagh 119, 125, 127
'Cuirasse-de-Fer' 23, 26
Curran Strand 49

Dalriada 59
Dalradian 2, 85
De Burgh 59
De Courcy 59
Devil's Port 35, 38, 56, 128, 134
Diatremes 38
Distillery 59
Dolerite 4, 30, 31, 38–40
Dooey 127
Downhill 27
Drumlin 94
Drumslade 127
Drumtullagh 44
Dry valleys 45, 49
Dunes 46, 48
Dundarave 126
Dungiven 42, 143
Dunluce 4, 16, 20, 24, 26, 40, 47, 51, 59, 60, 85, 99, 116, 121, 135
Dunluce Castle 38, 55, 59, 60, 116
Dunmull 4, 29, 129
Dunnaglea 13, 88, 90, 94, 123
Dunseverick 54, 125, 130
Dykes 3, 23, 40, 41, 138–139, 142

East Park 129
Elephant Rock 19
Entablature 27, 28, 124

INDEX

Eocene 3, 4, 24, 54
Erratics 43
Eskers 44
Estuarine alluvium 49
Estuarine clay 48

Farewell Skerry Member 76, 154
Faulting 42, 54
'Fish eyes' 7
Flandrian transgression 46
Flint 15–17, 44, 58, 85, 92, 93, 98, 99, 109, 110
Flow-banding 126, 127, 129, 136
Flow-till 45
Fluxion texture 130

Galboly Chalk 15, 16, 85–88, 90, 91
Gallery-grave 58
Garron Chalk 17, 86, 99–100
Geophysical logs 150–155
Geothermal gradient 148
Giant's Causeway 4, 5, 21–22, 26–28, 38, 46, 51, 55, 60, 68, 78, 123–124, 138
Giant's Cut 54, 102, 119, 132
Giant's Eyeglass 118
Gibbsite 26
Gigmagog's Grave 58
Girona 59
Glacial drainage channels 43, 45
Glacial striae 43, 44, 47
Glauconite 16, 90, 98
Glenarm Chalk 16, 85, 98
Glenmanus 49, 56
Glenstaghey 122–123
Glenstaghey Formation 6, 7, 74–77, 101, 155
Glentask 51, 60, 121, 125, 126
Goethite 140
Gortnabane 24, 56, 116, 135
Gravity surveys 42, 143
Great Gaw Fault 42
Great Stookan 28, 117, 118, 123
Greenland 23
Greenstone 4
Gulls Point 116
Guy Stone 126
Gypsum 7

Halite 7
Halloysite 26, 132
Hamilton's Seat 124
Hardgrounds 15, 85–87, 94, 98
Harryville 128
Hawks' Hollow 124
Hebridean Basin 15, 85
Hematite 51
Hettangian 10–14, 80
Hibernian Greensands 90, 102
Highland Border Ridge 2, 6, 15, 85
Highlandman's Bonnet 27, 123
Holywell Port 128
Hopefield 11, 39
Hornfels 9, 11, 136

Ice Age 3, 43
Iceland 23, 38
Iddingsite 140
Inchmearing 127
Inishowen 49, 54

Interbasaltic Bed 3, 5, 23, 26, 27, 56, 57, 60, 119–124, 132
Irish Harp 124
Iron Age 58
Iron ore 4, 5, 24, 26, 51, 60, 68, 119–124
Island Flockey 50
Islandbraddagh 50, 55, 125
Islandmore 26, 51, 60, 119, 125, 127
Island-tasserty 50
Island-varden 50
Islay 54, 58, 59

Jackson's Cove 55, 108, 135
Jan Mayen 23
Jointing 27, 28
Juniper Hill 44, 49, 127
Jura 54, 58
Jurassic 2, 3, 11–14, 78–83

Kaolinite 26, 132
Keevenagh 116
'Keuper' Marl 2, 6
Killygreen 44, 51, 60, 126
Kilmoyle 45
Kiltinny 42, 127
Kinbane 12, 38, 59
Kintyre 46, 58
Knocklayd 38
Knocksoghey 3, 40, 119, 137
Knocksoghey Formation 6, 7, 74–76, 154
Knocksoghey Sill 3, 23, 34, 39, 40, 137–138, 141

Lacada Point 118
Lacknamodeen 35, 117
Lagavarra Formation 6, 7, 74, 76, 155
Laminated Clays 43
Langford Lodge 29, 77
Lannimore Hill 45, 126
Lansdowne Crescent 9, 11, 36, 39, 68
Large Skerries 39, 40, 138, 140
Larry Bane Bay 16, 17, 23, 29, 31, 47, 55, 94, 95, 99, 133, 139
Larry Bane Chalk 16, 17, 85–87, 96
Larry Bane Head 16, 31, 33, 58, 59, 68, 90, 98, 99, 104, 105–107, 111
Laterite 4, 5, 23, 26, 28, 29, 40, 68, 119–124, 125, 132
Lava 3, 4, 23–29, 31
Leckilroy Cove 117, 118, 138
Leek Burn 50
Lemnagh 4, 26, 51, 60, 68, 78, 121–122, 132
Liassic 2, 4, 5, 11–14, 40, 54, 78–83, 85, 154
Lignite 23, 24, 26, 51, 60, 68, 119, 126
Limavady 6
Limestone 29
Lisburn 17
Lisnagunogue 42, 45, 121, 126
Listric surfaces 7
Lissanduff 58
Lisserluss 124
Lithomarge 26, 27, 40, 68, 119–124, 132
Little Ringan 127
Little Skerries 39, 40, 136, 140
Llanbedr (Mochras Farm) Borehole 13, 82

Long Gilbert Flint Band 17, 87, 110
Long Port 128, 134
Loom, The 27, 138
Lough Foyle 2, 54
Lough Neagh 58
Low Rock Castle 134
Lower Basalts 23, 24, 26, 42, 56, 116–119, 125, 130–131
Lower Sandstones 7

Maastrichtian 3, 15, 17, 85, 99, 110–111
Maddybenny 127
Magheraboy 121
Magheraclay 127
Magilligan 6, 8
Magmatic rolls 120
Magnetic anomalies 145
Manganese 118, 131
Mark Street 11, 39
Meadow Parks 49
Mercia Mudstone Group 6–8, 74–77
Mesolithic 58
Middle (Tholeiitic) Basalts 26–28, 124–126, 131
Mill Strand 47, 49
Mining 51, 60
Miocene 24
Miospores 76–77, 101
Montmorillonite 132
Moss-side 44, 50
Mount Druid 23, 118, 122
Mount Sandel 58
Moyarget 29, 45, 126, 130
Mugearite 23
Mulatto 15, 85
Mull 24, 54
Mullaghintorp 121
Murder Hole 118
Murlough Bay 6, 11, 42, 79, 90
Muschelkalk 6, 77
Myvatn 38

Nancy's Lope 121
Neolithic 58
Neptunists 4, 56
Normans 59
Norsemen 59
North Antrim Hardgrounds 16, 17, 85–87, 94, 98, 99

Old Castle Port 129, 131, 134
Old Red Sandstone 2
Oligocene 24
Olivine-basalt 5, 23–26, 29, 68
Opal 68
Organ 27, 55, 123–124
Out Hill 44, 129
Overflow Channels 43, 45
Oweynamuck 12, 16, 17, 68, 78, 88, 90, 102
Oweynamuck Flint Band 16, 17, 86, 91, 112

Pahoehoe lava 23, 24, 29, 116, 131
Palaeocene 3, 24
'Paramoudras' 17, 109
Passage-graves 58
Peat 3, 50, 51
Pegmatite 39, 40

Permeability 148
Permian 6
Permo-Triassic 6–8
Phreatic vents 38
Pigeonite 131
Plaiskin Head 40, 118, 124
Pleistocene 41, 43–45
Pliensbachian 10–14, 78
Polbrinck 138
Port Callan 125
Port Calliagh 111
Port Calliagh Chalk 111
Port Cool 128
Port Gallen 127–128, 134
Port Ganny 27, 28, 55, 118, 123–4
Port Gorm 46
Port Moon 55, 118, 124, 125, 132
Port More 2, 23, 26, 54, 109, 110, 111, 119
Port More Borehole 3, 6, 8, 11–14, 16, 39, 40, 75–77, 80–83, 85, 94, 98, 100, 111–113, 141, 143, 148–156
Port More Fault 16, 85, 123
Port More Formation 6, 7, 74–76
Port na Callian 118, 124
Port na Spaniagh 55, 118, 124, 138
Port na Tober 138
Port na Truin 118, 124
Port Noffer 27, 28, 55, 60, 117, 123–124, 138
Port Roestan 118
Portacallen 130
Portahapple 127
Portandoo 11, 138
Portaneevy 17, 29, 31, 99, 119, 133
Portballintrae 14, 26, 40, 43, 44, 45, 47, 49, 55, 58, 59, 80, 116, 138
Portbraddan 10, 16, 46, 55, 85, 88, 90, 93–97, 103, 130
Portcoon 117, 135, 138
Portcregcarragh 138
Portfad 124
Portnaboe 32, 40, 47, 118, 124, 142
Portnabrock 131
Portnacapple 47, 116
Portnahooagh 124
Portnakillew 13
Portnalea 116, 135
Portnalug 119
Portnawhellan 125
Portnee 136
Portninish 125
Portnool 99, 100, 108, 116, 135
Portrush 2, 3, 6, 9, 11, 24, 27, 36, 47, 49, 56, 58, 59, 79, 80, 85
Portrush Chalk 15, 17, 86, 100, 109
Portrush Fault 16, 24, 29, 42, 50, 85, 119, 125
Portrush Rock 4, 9, 11, 56
Portrush Sill 3, 4, 5, 9, 11, 12, 23, 38, 39, 40, 54, 135–137, 139, 145
Portscadden 39, 136, 140, 141
Portstewart 29, 38, 40, 44–49, 56, 58, 68, 127–128
Portstewart Point 128
Pottagh Bridge 49
Pound, The 38, 55, 116, 135

Priest's Hole 135
Promontory forts 58
Pyrite 11

Quarrying 51
Quartz 44, 68
Quaternary 3, 43–50

Raised Beaches 46, 54–56, 128
Ramore Head 36, 39, 47, 136, 145
Rathlin Island 11, 24, 26, 46, 93, 116, 131, 143
Raths 58
Retreat Castle 96
Reviggerly 11, 39, 40, 139, 140
Rhaetic 2, 6, 8, 40, 154
Rhyolite 23
Riggin, The 56, 68, 99, 108, 135
Ringree Point 134
Ripple marks 7
River terrace deposits 49
Roches moutonnées 43, 44
Rock House 127
Rock Ryan 135
Rockview 127
Roveran Valley Head 32, 40, 55, 118, 123–124, 138, 142
Runkerry Point 16, 42, 55

St Thomas' Cliffs 97
Salt marsh 49
Sand and gravel 44, 57
Santonian 15–17, 85
Schist 2, 44
Scotland 23
Scottish Ice 3
Scottish Readvance 43, 46
Scrabo 38
Scudion Craig 29, 57, 125, 129
Sea Gull Isle 118
Sedimentary structures 7
Seismic velocity 148
Senonian 3, 15–17
Sheelas Head 17
Sheep Island 2, 23, 39, 40, 54, 55
Sherwood Sandstone Group 6, 7, 76, 148, 155
Silica 3, 39, 40
Sills 135–138, 139
Sinemurian 10–14
Sixteen-Inch Skerry 154
Skerries, The 2, 37, 39, 56, 136–137
Slidderycove 97–99
Slimag 44
Solifluxion 45
Sonic logs 150–154
South Antrim Hardgrounds 98
Spanish Organ 124
Spaniard Rock 118
Sperrin Mountains 2
Spillways 43, 45
Stackaboy 55
Stackahorlin 31, 34, 133
Stackandoon 119
Stacks 46, 54, 56
Stony Port 128
Storks, The 39, 40
Stradley Rock 135

Strand lines 46
Strandhead 47, 48, 56, 127
Striated pavements 43, 44
Stromatolites 15
Submerged peat 47
Sugarloaf 116
Suitcase Sandstone 6, 76
Sycamore Port 128

Tachylite 121, 140
Tanderagee Chalk 15, 17, 87, 110–111
Tea-Green Marls 8
Templastragh 46
Tertiary 3, 23–41
Tholeiitic basalt 3, 5, 23, 26–28, 57, 58, 124–126, 131
Till 44, 45
Three-foot Skerry 6, 76
Tonduff 24, 124
Toome 58
Torr Head 2, 46
Tow Valley Fault 145
Trachyte 23
Tree-moulds 24
Triassic 2, 6–8, 74–77
Trostan 29
Tubber Patrick 127
Tuff 3, 23, 24, 26, 31, 38, 131, 118–119, 133
Tullaghmurry 48, 49
Turfahun 130

Ulster Museum 12–13, 80
Upper Basalt 24, 29, 127–130
Upper Boreal Age 48
Upper Glauconitic Sandstone 90
Upper Interbasaltic Bed 24, 28–29, 60, 126
Upper Sandstone 7
Urbalreagh 51, 58, 60, 121

Vents 5, 23, 29, 31, 38, 116, 128, 133–135, 139
Victoria Jubilee Bridge 55
Volcanic ash 3, 23, 31, 56
Volcanic plugs 38, 39, 135, 139
Vulcanists 4, 56

Walk Mills 59, 126
'Warp' 48
Warren Lodge 48
Wash Tub 39, 141
Water supply 51
Weir's Snout 123, 124
West Park 60, 129
Westbury Formation 8, 14
Weybourne Chalk 17
White Limestone (Chalk) 3, 15–19, 44, 51, 68, 84–115
White Park Bay 10–14, 16, 42, 45, 46, 47, 49, 78, 80, 83, 88, 91, 93, 103, 105, 124, 139
White Rocks 2, 5, 16, 17, 35, 38, 47, 55, 56, 98, 99, 108, 116, 134, 138
Whitehead Flint Band 94
Winkle Island 39, 40, 137, 138, 140

Xenoliths 5, 12, 39

Zeolites 5, 24, 39, 40, 68, 130, 139

HER MAJESTY'S STATIONERY OFFICE

Government Bookshops

80 Chichester Street, Belfast BT1 4JY
49 High Holborn, London WC1V 6HB
13a Castle Street, Edinburgh EH2 3AR
41 The Hayes, Cardiff CF1 1JW
Brazennose Street, Manchester M60 8AS
Southey House, Wine Street, Bristol BS1 2BQ
258 Broad Street, Birmingham B1 2HE

*Government publications are also
available through booksellers*

INSTITUTE OF GEOLOGICAL SCIENCES

20 College Gardens, Belfast BT9 6BS

Exhibition Road, London SW7 2DE

Murchison House, West Mains Road,
Edinburgh EH9 3LA

The full range of Institute publications is
displayed and sold at the Institute's Bookshop
at the Geological Museum, Exhibition Road,
London SW7 2DE

*The Institute was formed by the incorporation of
the Geological Survey of Great Britain and the Geological
Museum with Overseas Geological Surveys and is a
constituent body of the Natural Environment
Research Council. The Geological Survey of
Northern Ireland is an agency service provided by the
Institute of Geological Sciences to the Department
of Commerce, Northern Ireland*

Printed in Northern Ireland for Her Majesty's Stationery Office by
W. & G. Baird, Ltd., Antrim.

Dd. 075850. 3/78. K8. W.&G.B. 55-5079.